Introduction to China's S&T Innovation

Introduction to China's S&T Innovation focuses on China's economic development, healthcare, digital construction and other hot spots of people's livelihood, and showcases the fruits of reform and development that have benefited all Chinese people.

In recent decades, with the wave of economic globalization, China's economy has taken off rapidly, and at the same time, China has also made great achievements in science and technology, education and culture, which have attracted the world's attention. This series is a classic work that helps overseas readers understand China's economic and social development in all aspects. Each book in the series is edited by authoritative experts and scholars in the relevant fields. By elaborating on the development history, technological innovation, policy evolution, underlying logic and international cooperation in their respective fields, the series comprehensively introduces the quality and speed of China's modernization, and shares the experience of Chinese-style modernization, which will help the world's people to better understand China, and to promote the exchanges and mutual understanding of different civilizations, and promote common development.

Wei Liu · Wenbo Xu · Xiaofeng Zhao ·
Yunqing Cheng
Editors

Digital China

Digital Industries, Industry Digitalization and Digital Society

華中科技大學出版社
Huazhong University of Science and Technology Press

Springer

Editors
Wei Liu
School of Electronic Information and Communications
Huazhong University of Science and Technology
Wuhan, Hubei, China

Xiaofeng Zhao
School of International Education
Huazhong University of Science and Technology
Wuhan, Hubei, China

Wenbo Xu
School of Navigation
Wuhan University of Technology
Wuhan, Hubei, China

Yunqing Cheng
School of Electronic Information and Communications
Huazhong University of Science and Technology
Wuhan, Hubei, China

Introduction to China's S&T Innovation
ISBN 978-981-95-6239-8 ISBN 978-981-95-6240-4 (eBook)
https://doi.org/10.1007/978-981-95-6240-4

Jointly published with Huazhong University of Science and Technology Press
The print edition is not for sale in China (Mainland). Customers from China (Mainland) please order the print book from: Huazhong University of Science and Technology Press.

This Springer imprint is published by the registered company Springer Nature Singapore Pte Ltd.
The registered company address is: 152 Beach Road, #21-01/04 Gateway East, Singapore 189721, Singapore

Series Editor's Preface

Throughout the history of mankind, when education in a country thrives, the country will thrive, and strong education makes a strong nation. Today, as economic globalization continues to deepen, the opening-up of education has become an essential strategy for national development. Since the 18th National Congress of the Communist Party of China, numerous important documents concerning the internationalization of higher education have been issued, with international exchange and cooperation now being one of the five major functions of universities. Huazhong University of Science and Technology (HUST) has actively responded to this call by releasing a series of guiding documents to promote the internationalization of education, and placing significant emphasis on the education of international students in China. The "Introduction to China's S&T Innovation" is a classic collection of works created for international students in China in this context, which can be said to be both a need of the times and a responsibility.

We live in an era characterized by advanced transportation, frequent interpersonal exchanges, and increasingly dynamic cultural interactions. In this era, how to tell China's stories well and let the world understand China better has become an important issue for us. We are fully aware that international students in China possess the unique advantage of being "bridges between China and the world," serving as special envoys of friendship between China and the world and important connectors between China and the world. By providing international students with a comprehensive understanding of China's economic and social development, we can empower them to tell China's stories well, thereby enhancing the international dissemination of Chinese civilization.

In recent years, Xi Jinping, General Secretary of the Central Committee of the Communist Party of China, has written numerous heartfelt letters to overseas students, returning scholars, and international students in China. Studying these letters has deeply inspired and encouraged me. On the one hand, as a former overseas student who studied abroad in Germany and the United States for over six years starting in 1995, I have gained a deep understanding of the academic and living needs of international students in China. On the other hand, since returning from studying abroad over 20 years ago, I have been engaged in research, teaching, and administrative work at universities. During my tenure as Vice President of Huazhong University of Science and Technology, I was responsible for international exchange and cooperation for many years, with international student education being one of my primary responsibilities. I have witnessed and experienced the rapid development of China's higher education and scientific research. As the number of international students in China continues to grow, I feel proud and honored, and at the same time, I also feel a great sense of responsibility.

At present, China is in a new phase of high-quality development, where promoting high-quality development through high-level opening-up has become a requirement of the times, both in the economic and educational fields. I believe that high-level opening-up of education is not only a driving force for the "Double First-Class" initiative in Chinese universities, but also an opportunity to explore new areas of international cooperation for universities. Therefore, in our internationalization efforts, we have consistently adopted an international perspective, positioning ourselves within a global framework, benchmarking against world-class universities, and actively expanding substantial cooperation with top international institutions. While expanding the scale of international students in China, we have also worked to further improve the quality of their education.

The "Introduction to China's S&T Innovation" selects key topics related to China's economic development, medical and health care, digitalization construction and other hot spots of China's livelihood. By explaining the development process, technological innovations, policy evolution, underlying logic, and international cooperation in these fields, the series provides a comprehensive introduction to the quality and speed of China's modernization efforts. It not only serves as a vivid record of China's modernization, but also highlights the spiritual symbols and cultural essence of Chinese civilization.

Each book in this series is edited by authoritative experts in the relevant field. Here, I would like to extend my heartfelt gratitude to the chief editors for their strong support and hard work. They are leading scholars deeply engaged in scientific research, with keen insights into the trends of economic and technological development in China and internationally. They possess rich experience in book compilation and a profound understanding of the content. Furthermore, they are front-line educators with extensive teaching experience, who understand the needs of international students in China, know the ways of teaching, and have unique insights into talent cultivation. I am confident that this series, which fits the background of the times, carefully selects hot topics, and understands readers' needs, will be able to present a comprehensive, multidimensional, and authentic China to international students. It will empower them to tell a good story about China, act as good envoys of friendship, and build bridges of cooperation.

Looking ahead, the "Introduction to China's S&T Innovation" will continue to be developed and published, laying a solid foundation for the education of international students in China. We will continue to advance the internationalization of our university step by step. I look forward to seeing international students and overseas readers get to know, understand, and fall in love with China through the "Introduction to

China's S&T Innovation." I also hope that this series will become a beautiful landscape of cultural exchange between China and the world, contributing our wisdom and strength to the building of a community with a shared future for mankind.

Jianguo Chen
Former Vice President
Huazhong University of Science
and Technology
Wuhan, Hubei, China

Preface

In the era of deep integration of economic globalization and digital informatization, a new generation of information and communications technology (ICT), represented by artificial intelligence, big data, cloud computing, blockchain, etc., is reshaping the global economic landscape and revolutionizing the social operation modes with unprecedented momentum. These technologies have deeply penetrated into various industries, giving rise to new business models and forms, and driving profound changes in the global economic landscape. Major economies in the world, such as the United States, Russia, the European Union, and Japan, have keenly perceived this development trend and formulated their digital development goals for 2030 across diverse domains.

China has made remarkable achievements in the development of the information industry and information infrastructure construction, attracting worldwide attention and emerging as a global leader in digital transformation. In February 2023, China released the "Overall Layout Plan for Digital China Construction", clearly outlining the ambitious goal of elevating its digital development level to the forefront of the world by 2035. Currently, a new wave of digitization is sweeping across China, which is about to trigger profound changes in society, economy, production, and life, and reshape China's future development pattern.

Against this background, it is imperative to update the content of information literacy general courses in colleges and universities. This is conducive to helping undergraduate and postgraduate students broaden their digital horizons and keep up with the digital development trend. However, compared with other general courses, there is a lack of books in China that comprehensively introduce "Digital China". International students studying in China show great enthusiasm for China's digital development. Yet currently, they mostly get a partial understanding through digital applications in their daily lives. This textbook, Digital China (Bilingual version), is thus written to meet this need. Comprising eight chapters, the book covers topics such as the basics of the ICT industry, overview of "Digital China", ICT innovations, digital industrialization, industrial digitalization, digital society, cyberspace governance, and international digital cooperation, aiming to present readers with a panoramic view of "Digital China".

Chapter 1, "Basics of the ICT Industry", primarily introduces the basic concepts and features of the ICT industry. This chapter first traces the development trajectory of the ICT industry, from the birth of language and writing, the rise of postal services, telecommunications, and broadcasting, to the revolutionary changes in the computer and Internet era. Then it deeply analyzes the features of the ICT industry, including its industrial scope, position in the economic system, inherent features, and development laws. Subsequently, it introduces the new-generation information technologies and the Fourth Industrial Revolution, and explores the changes and opportunities they bring to the ICT industry. Finally, it focuses on the new digital economy, interpreting the covered areas and the related digital strategic layout made by different countries.

Chapter 2, "Overview of Digital China", introduces the basic background and existing plans of China's ICT industry. Firstly, this chapter briefly reviews the development journey of China's ICT industry, from its humble beginnings to its current strength, in chronological order. Then it expounds on the current scale and development level of Digital China, enabling readers to have a clear understanding of its status quo. Subsequently, it analyzes related macro strategies and industrial plans of China's ICT industry, covering national development strategies, government plans, and policy tools. Finally, it focuses on the development goals and construction priorities of "Digital China", introducing the Chinese government's development plans in digital infrastructure, digital economy, digital society, and other aspects.

Chapter 3, "ICT Innovations", interprets how China promotes the integrated development of information technology, and industrial innovation with the national economy through overall planning. This chapter first introduces the composition of China's scientific research and development system, as well as the organization and implementation of scientific and technological development plans. Then, through case studies, it introduces major scientific and technological infrastructure projects such as supercomputers, quantum communications, and BeiDou navigation, along with scientific and technological achievements in industrial research and development such as Ultra-HD video, mobile communications, and integrated circuits.

Chapter 4, "Digital Industrialization", elaborates on the main achievements of China's traditional information industry in the early stage of the "14th Five-Year Plan". This chapter successively introduces four digital industry sectors: basic telecommunications industry, Internet industry, software and information technology services industry, and electronic information manufacturing industry. In the aspect of basic telecommunications industry, it focuses on the construction progress of new-type information infrastructure in China, such as 5G mobile communications networks, satellite Internet, Internet of Vehicles, and computing infrastructure. In the aspect of Internet industry, it focuses on the achievements of China in characteristic applications such as social media, e-commerce, and mobile payments. In the aspect of software and information technology services, it introduces the development of cloud computing, big data, artificial intelligence, and blockchain in recent years. In the aspect of electronic information manufacturing industry, it emphasizes on the technological breakthroughs made by China in fields such as new display devices and integrated circuit chips.

Chapter 5, "Industrial Digitalization", presents China's representative achievements in promoting the digital transformation of industries such as manufacturing, agriculture, energy, and transportation by leveraging informatization to drive industrialization. This chapter first focuses on intelligent manufacturing, showing the digital transformation of the manufacturing industry from the industrial Internet platforms that build industry-wide connections, to the use of industrial robots to improve production efficiency, and then to the intelligent upgrading of production in intelligent factories. Then it pays attention to smart agriculture, introducing how agricultural big data helps with scientific decision-making and how smart cultivation and smart farming promote the modernization of agriculture. Subsequently, it explores the field of smart energy, covering the construction achievements of intelligent coal mines, smart power plants, and smart power grids. Finally, it enters the field of intelligent transportation, showing the high-efficiency and convenience of smart road transportation, intelligent rail transportation, and intelligent water transportation.

Chapter 6, "Digital Society", tells China's practice of using digital technologies to improve the level of public services in cities and rural areas and provide better social services such as education and healthcare. This chapter first introduces China's exploration in smart cities, from the coordination of urban traffic, the efficient operation of the urban brain, to the innovative construction of digital twin cities, showing China's measures to improve urban governance with digital technologies. Then it introduces various measures taken by China to promote the construction of digital villages, such as strengthening the construction of communications infrastructure to assist the "poverty alleviation" campaign, establishing a complete rural digital public service system, and supporting the development of rural e-commerce to achieve rural revitalization. In the section on smart education, it covers diverse applications such as the construction of digital campus facilities, the establishment of digital course platforms, and Internet-based education. The smart healthcare section presents the progress of digital healthcare infrastructure and the practical application results of Internet healthcare services and smart healthcare applications.

Chapter 7, "Cyberspace Governance", elaborates on the current situation and achievements of China in the field of cyberspace governance. This chapter first expounds on the basic concept of cyberspace and the needs and challenges faced in security and governance. Then it introduces the legal framework of cyberspace constructed by China, covering legal construction in multiple aspects, including the protection of personal rights and interests, digital economic rules, network information security, and network ecological norms. Then it introduces the achievements of China in the construction of cyberspace security protection capabilities, including cybersecurity technical standards, the cybersecurity protection forces, and the development of the cybersecurity industry. Finally, it introduces the comprehensive and multi-dimensional governance carried out by China in cyberspace social governance, involving public environment governance, personal privacy protection, digital content supervision, and digital economic regulation.

Chapter 8, "International Digital Cooperation", introduces China's contributions to global cyberspace governance. This chapter first points out the challenges and opportunities faced by global digital cooperation in cyberspace, including cybersecurity risks and digital trade demands. Subsequently, it expounds on China's proposition of building a global "community with a shared future in cyberspace", including its core ideas and profound connotations. Then it introduces China's achievements in the development of the international digital economy, especially in the fields of global digital service supply and cross-border e-commerce. Finally, in the part of international cooperation on cyberspace governance, it introduces the international cooperation carried out by China in cybersecurity cooperation and the fight against cybercrime.

Chinese President Xi Jinping sent congratulations to 2022 World Internet Conference Wuzhen Summit and stressed: "China is willing to work with countries around the world to jointly explore a path of global digital development featuring the joint construction and sharing of digital resources, the vibrant development of the digital economy, well-targeted and efficient digital governance, the prosperous development of digital culture, strong digital security protection, and mutually beneficial digital cooperation. We aim to accelerate the building of a community with a shared future in cyberspace and contribute wisdom and strength to world peace and development and the progress of human civilization." Digital China (Bilingual version) serves as a communication bridge between China and the world. It provides a window for international students to gain in-depth knowledge of China's digital development, and enables them to experience the innovative vitality and development achievements of China's digital field, thus contributing wisdom and strength to future international cooperation and exchanges.

The writing team of this book is composed of the faculties engaged in international student education from the School of International Education and the School of Electronic Information and Communications at Huazhong University of Science and Technology (HUST). Liu Wei, Xu Wenbo and Cheng Yunqing are responsible for organizing and writing the chapters of basics of the ICT industry, overview of Digital China, ICT innovations, digital industrialization, and industrial digitalization. Zhao Xiaofeng and Ma Rong are responsible for organizing and writing the chapters of digital society, cyberspace governance, and international digital cooperation. With the support of International Education Base of Engineering at HUST, some of the book's content was taught in the form of courses and lectures and revised based on the opinions collected from international students. This writing process has also received support from the project "Promoting International Student Education Innovation through Digital Education Development" (23LH0102) by China Association of Higher Education.

Digital China is a grand project of comprehensively empowering the high-quality development of the economy and society through digitalization that China is currently vigorously promoting. It covers a wide range of areas, involves complex content, and its construction progress is changing rapidly. The eight-chapter content of this book is only a microcosm of contemporary China's digital construction. In order to provide a panoramic introduction within a limited space, the writing team mainly referred to publicly available materials released by government departments and industry associations since the "14th Five-Year Plan". Although the writing team has gone all out, due to limitations in professional expertise, time, and energy, there are bound to be some oversights in this book. We sincerely welcome readers to offer their criticisms and suggestions.

Wuhan, China

Wei Liu
Wenbo Xu
Xiaofeng Zhao
Yunqing Cheng

Contents

List of Figures

List of Tables

Chapter 1
Basics of the ICT Industry

1.1 Brief History

Information is the representation of the movement and changes of things. From the invention of language and script by the ancients to written communication, and futher to the use of electronic technology to process and disseminate information, the history of human information exchange has spanned thousands of years. Human beings have experienced five innovations in information technology, and a corresponding information industry has been formed in its promotion and application.

1.1.1 Language and Script

The emergence and use of language and script are generally regarded as the beginning of human information technology innovations, which have promoted the development of ancient agriculture, handicrafts, and commerce. Human language is considered as the most primal and enduring conduit of information. It stands as the paramount instrument of interpersonal communication, and a symbolic medium through which expressions are conveyed. Script, as the written symbol carrying the mantle of language, is a tool used by humans to convey information, enabling the transmission of information to transcend the boundaries of time and space for posterity.

Modern Homo sapiens, endowed with robust linguistic communication abilities, embarked on a momentous journey "out of Africa", venturing into the vast expanse of the Eurasian continent. Over the course of millennia, diverse ethnic groups across the globe cultivated their unique languages, among which the four major ancient civilizations developed their own distinctive scripts, as shown in Table 1.1.

The shape of script is closely related to writing methods and record carriers, and gradually evolves with the development of social productivity. The following takes

W. Liu et al. (eds.), *Digital China*, Introduction to China's S&T Innovation,
https://doi.org/10.1007/978-981-95-6240-4_1

Table 1.1 Overview of the four ancient civilizations

Civilizations	Appearance time	Geographic regions	Scripts	Examples
Egyptian Civilization	About 3000 BC	Nile River Basin in Africa	Hieroglyph	
Indus Valley Civilization	About 2500 BC	Indus Valley in Asia	Seal script	
Chinese Civilization	About 2070 BC	Yellow River Basin and Yangtze River Basin in Asia	Oracle bone script	
Babylonian Civilization	About 1800 BC	Tigris and Euphrates River Valleys in Asia	Cuneiform	

the evolution of Chinese script for case study. The earliest Chinese script with the characteristics of written symbols that can be traced back are the written symbols engraved on animal bones in the remains of the Longshan Culture around 2500 BC. This type of written symbol has the characteristics of hieroglyphic script. The oracle bone script that appeared in the Shang Dynasty around 1400 BC is regarded as the first form of Chinese script. It was engraved on tortoise shells and animal bones for the purpose of divination. The Chinese calligraphy brush pen that appeared in the Zhou Dynasty can be used to write characters on bamboo slips or textiles (such as silk). Different vassal states at that time had their own written scripts, which obviously hindered the cultural communication in ancient China.

In 221 BC, China's first emperor Qin Shihuang unified China and declared "small seal script" as the standardized national script, marking the first formalization of Chinese script writing. During the Eastern Han Dynasty, around 105 AD, paper-making technology was improved and paper began to be used as the main writing material; at the same time, clerical script and standard script appeared, which were simpler and easier to recognize. After the development of the Han Dynasty, the

Chinese script system reached maturity and the shapes almost ceased to evolve, so Chinese character is also called "Hanzi". The number of Chinese characters increased over time. By the Qing Dynasty, the Kangxi Dictionary had collected 47,035 Chinese characters.

Traditional Chinese characters have many strokes, which makes them difficult to read and remember. In 1956, the State Council of the People's Republic of China promulgated the "Scheme for Simplifying Chinese Characters". In 1964, the Chinese Script Reform Committee officially compiled and printed the "General Table of Simplified Chinese Characters", which includes 2238 simplified Chinese characters. They are the currently used Chinese characters. Table 1.2 takes three Chinese characters as examples, and lists the evolution processes of their form.

The emergence of script makes it possible to record knowledge and history. All dynasties of China attached great importance to the recording of history. The Records of the Grand Historian, written between 104 and 91 BC, documents Chinese history from ancient times up to the Han Dynasty. Among the four ancient civilizations in

Table 1.2 Examples of Chinese character evolution

Fonts	Appearance periods	Recording methods	Example 1 (horse)	Example 2 (fish)	Example 3 (volume)
Oracle bone script	About 1400 BC	Engraved on tortoise shells or animal bones			
Bronze inscription	About 1300 BC	Cast or engraved on bronze			
Small seal script	221 BC	Written on bamboo slips			
Clerical script	About 150 AD	Written on paper	馬	魚	冊
Standard script	About 220 AD	Written on paper	馬	魚	冊
Simplified Chinese Character	1955	Written or printed on paper	马	鱼	册

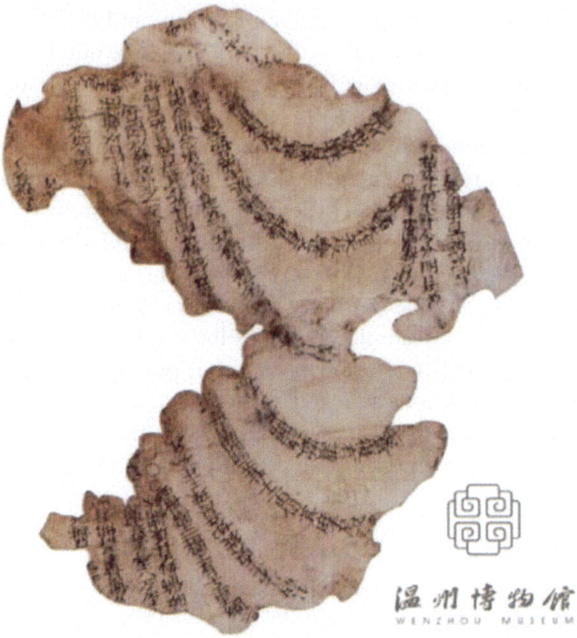

Fig. 1.1 Fragments of an ancient book printed by movable type printing (Northern Song Dynasty, about 1100 AD)[1]. *Image Source* Wenzhou Museum

the world, Chinese Civilization is the only one that has continued to this day without interruption and has complete written historical records.

1.1.2 Printing and Postal Service

There have been many kinds of carriers for recording texts in human history, and paper is the most convenient and widely used carrier. Before the emergence of printing, humans recorded texts by hand-written transcription, with limited scope and scale of application. In the Northern Song Dynasty, around 1040 AD, movable type printing emerged, making it possible to reproduce books on a large scale. One example is as shown in Fig. 1.1. Movable type printing was spread to Korea Peninsula, Japan and Central Asia, and then to Europe through Arabia and other regions, promoting the development of world civilization.

In ancient times, the exchange and dissemination of information mainly relied on the transportation of paper carriers. Ancient China established a system dedicated to the transportation of information carriers and materials, called "Youyi", in which "You" were the institutions dedicated to the delivery of official documents, and "Yi"

[1] Fragments of pages from the movable type printing of *The Sutra on the Contemplation of Amitabha Buddha*, dating back to the Northern Song Dynasty, are now in the collection of the Wenzhou Museum in China. They are the earliest movable type prints discovered so far.

Fig. 1.2 The image of a postman in Youyi system (Wei Jin Periods, about 220–420 AD). *Image Source* Gansu Provincial Museum

were the transfer stations for the horses and vehicles that delivered those documents. Ancient dynasties usually set up specialized stations called "Yizhan" in major transportation towns, and the documents were relayed from one station to another. In the Han Dynasty, urgent documents could be delivered 400 miles (about 200 km) in one day. Figure 1.2 shows a painted brick of a postman unearthed in Jiayuguan on the Silk Road in the Wei Jin Periods. The Youyi system was mainly for official services and was not abolished until the establishment of the postal system in the Qing Dynasty in 1896, existing for about 2000 years.

With the development of society, an industry that specializes in providing letter and parcel transportation services, namely the postal industry, emerged. In the Ming and Qing Dynasties in China, there were "Xinju" that provided mailing services to the people; however, the delivery services of various Xinju were not interoperable. The modern postal industry was born in the United Kingdom (UK). In 1840, the UK issued the world's first stamp, the "Penny Black", and promoted the development of the postal industry through prepaid postage. In 1874, the "General Postal Union" (now the Universal Postal Union) was established, and sending letters through the postal system became the main way for countries around the world to exchange information at that time.

1.1.3 Telecommunications and Broadcasting

From the mid-nineteenth century to the mid-twentieth century, electromagnetic theory gradually matured, and it was possible to make electrical signals as a carrier for communication. This brought fundamental and great evolution for the communication in human lives. The transmission of information can break away from conventional audio-visual method and use metal wires, or even rely on invisible and intangible electromagnetic waves.

American scientist Morse invented the telegraph, which transmitted English letters and other texts as a combination of dots and dashes. By sending a combination of electrical signals of different lengths, telegraph realized the transmission of text information. American inventor Bell invented the telephone, which could convert voice into electrical signals through microphones and speakers, thus realizing the transmission of language information through cables. Italian scientist Marconi invented the wireless telegraph, using air as the medium. The radios and televisions developed by British and American scientists can receive broadcasting signals from radio and television stations, then demodulate, recover and play the audio and video signals. The birth of telegraph, telephone, radio and television have changed the traditional way of communications that relied on paper media, and realized cross-regional long-distance communications. Table 1.3 lists the major inventions in the fields of telecommunications and broadcasting.

Telegraph, telephone and other communications technologies have also given birth to modern telecommunications services, with companies such as the American Telephone and Telegraph Company (AT&T) serving as examples. Meanwhile, radio broadcasting and television have given rise to modern media services, exemplified by orgnizations like the British Broadcasting Corporation (BBC). The United Nations (UN) first formulated the "International Standard Industrial Classification of All Economic Activities" in 1949, using code 73 to refer to communications, which includes not only the traditional postal industry, but also the information transmission services based on wired and wireless technologies, as well as the related information exchange, storage and recording services.

Table 1.3 Major inventions in the fields of telecommunications and broadcasting

Fields	Inventions	Representative events
Telecommunications	Cable telegraph	In 1837, American scientist Morse successfully developed the electromagnetic telegraph based on wired cables; in 1858, the submarine telegraph cable connecting Europe and America across the Atlantic Ocean was successfully laid
	Wireless telegraph	In 1896, Italian scientist Marconi successfully developed the wireless telegraph, and in 1901, transatlantic radio communication was realized
	Wired telephone	In 1876, American inventor Bell applied for a patent for the telephone he invented; in 1878, the long-distance telephone call experiment between Boston and New York was successfully done
	Telephone exchange	In 1926, the world's first large-scale crossbar automatic telephone exchange was put into use in Sweden; in 1938, the United States (US) deployed a crossbar automatic telephone switching system
Broadcasting	Electron tube	In 1904, British scientist Fleming invented the vacuum diode; in 1906, American inventor Lee de Forest invented the vacuum triode; in the late 1930s, electron tube became the mainstream electronic component
	Radio	In 1906, American scientist Fessenden conducted a broadcasting experiment using modulated radio waves to send music and speech; in 1910, Americans Dunwoody and Piccard successfully developed the crystal radio based on electron tubes
	Television	In 1933, American scientist Zoelkin successfully developed the camera tube and image tube for television; in 1954, the US successfully produced the world's first color TV

1.1.4 Computer and Internet

During the more than 50 years from the mid-twentieth century to the early twenty-first century, breakthroughs were made in technologies such as semiconductors, computers, the Internet, and mobile communications, setting off a wave of information technology revolution.

The development of microelectronics provides a theoretical basis for the technological upgrading of electronic components, devices and circuits. In 1958, the invention of integrated circuit (IC) put the development of semiconductor technology on the fast track. In 1965, Gordon Moore predicted that the number of transistors on a chip would double every 18 months in the future. The rapid development of the semiconductor industry and technology has laid a matarial foundation for the production and manufacturing of various electronic information products. Table 1.4 lists the major technological advances in the field of microelectronics.

Table 1.4 Major technological advances in the field of microelectronics

Achievements	Representative events
Transistor	In June 1948, the solid-state physics research group at Bell Laboratories invented the transistor; in 1953, Shockley and others from the research group successfully developed the silicon junction field-effect transistor; with its small size and stable performance, the transistor began to replace the electron tube as the basic device of electronic information products
Integrated circuit	In September 1958, Kilby of Texas Instruments developed the world's first germanium integrated circuit board containing transistors, resistors and capacitors; in 1959, Noyce of Fairchild Semiconductor used planar technology to produce silicon integrated circuits, truly realizing monolithic integrated circuits
Large scale integrated circuit (LSIC)	In 1967, Bell Laboratories in the US produced the world's first large-scale integrated circuit (integrating 1 000 components on a single chip), which was soon promoted to industrial production and practical applications

Theoretical breakthroughs have been made in the principles and architecture of modern computers. In 1945, scientist John von Neumann proposed the scheme of general-purpose electronic computer with binary storage program. In 1946, the first electronic computer was successfully created in the US. With the development of microelectronics, computers have entered the era of integrated circuit from the era of electron tubes and transistors, and computing power has been greatly improved with the improvement of the production process of central processing unit (CPU). Computers have changed from large equipment that only a few institutions have the opportunity to use to consumer electronic devices that have entered thousands of households. Table 1.5 lists the major technological advances in the field of computer industry.

The global Internet for computer interconnection has been verified and applied. With the funding of the Defense Advanced Research Projects Agency (DARPA) of the US Department of Defense, major scientific research institutions in the United States gradually connected to the ARPANET (also known as the Advanced Research Projects Agency Network) during a period of 20 years, and explored solutions suitable for large-scale heterogeneous network interconnection. In 1995, after the Internet began commercial operation, it quickly became popular around the world and became the most mainstream communication network. Internet information services such as the World Wide Web, search engines, online e-commerce, and social networks have brought the world into the Internet age. Table 1.6 lists the major technological advances in the field of Internet.

Public cellular mobile communications networks provide global roaming wireless services. In the 1980s, the world began to deploy personal mobile communications networks based on cellular base stations. The demand for mobile communications gradually shifted from voice services to data services, and mobile phones were widely used to send and receive text messages and emails. In 2007, Apple Inc. launched the

Table 1.5 Major technological advances in the field of computer industry

Technologies	Representative events
Vacuum tube computer	In February 1946, Professors Mauchly and Eckert of the University of Pennsylvania successfully developed the computer ENIAC, which was the world's first general-purpose computer, using about 18,000 electron tubes, and was used for national defense and scientific calculations; in 1951, the computer UNIVAC developed by Remington Rand Corporation was used in the US census, and computers began to serve civilians
Transistor computer	In 1954, Bell Laboratories in the US developed the world's first transistor computer, TRADIC, while at the same time, operating systems and algorithmic languages were born, and high-level languages such as COBOL and LISP entered the practical stage
Integrated circuit computer	In April 1964, the International Business Machines Corporation (IBM) launched the world's first computer series, the IBM System/360 using integrated circuits; in 1968, IBM released the world's first commercial database system, the Information Management System (IMS); in 1981, IBM launched the world's first personal computer, the IBM 5150, and proposed an architecture standard for using microprocessors
Computer CPU	In 1971, Intel Corporation of the US released the world's first 4-bit microprocessor 4004 (containing more than 2000 transistors and produced with a 10-micron process); in 1979, Intel Corporation launched the 8-bit microprocessor Intel 8088 (containing about 29,000 transistors); in 1985, Intel Corporation launched the 32-bit microprocessor 80,386 (containing about 275,000 transistors); in 1993, Intel Corporation launched the Pentium processor (containing about 3 million transistors and produced with a 0.8-micron process); in 1999, Intel Corporation launched the Pentium III processor (containing about 9.5 million transistors and produced with a 0.25-micron process); in 2006, Intel Corporation launched the Itanium processor (containing 1.72 billion transistors and produced with a 90-nm process)
Computer software	In 1970, Bell Laboratories in the US launched the Unix operating system; in 1970, Edgar Frank Codd of IBM proposed the relational database theory; in 1984, Apple Computer launched the computer Macintosh (Mac for short), which provided a graphical operating system interface; in 1991, an university student from Finnish, Linus Torvalds, developed the open source Linux operating system; in 1995, Microsoft Corporation of the United States released the Windows 95 operating system and the office software Microsoft Office 95, which became the mainstream computer software

iPhone, marking the arrival of the era of touch-based smartphone. Mobile phones began to gradually replace personal computer (PC) as the main electronic device for personal Internet access, and applications running on mobile phones became the main form of personal software. Table 1.7 lists the major technological advances in the field of mobile communications.

Information and communications technology (ICT) is represented by semiconductors, computers, the Internet, mobile communications and other technologies. It is based on microelectronic and optoelectronic technologies, supported by computers and communications technologies, and covers the acquisition, transmission, storage,

Table 1.6 Major technological advances in the field of Internet

Technologies	Representative events
ARPANET	In 1966, the Defense Advanced Research Projects Agency of the US Department of Defense launched the ARPANET research project; in 1969, the University of California, Los Angeles and the Stanford Research Institute achieved the first long-distance communication on a packet switching network; in 1974, the first version of the Transmission Control Protocol (TCP) for heterogeneous network computer networking communications was officially released; in 1983, TCP was decomposed into the TCP/IP protocol suite and used by the ARPANET
Internet	In 1989, Web services based on Hypertext Markup Language (HTML) and Hypertext Transfer Protocol (HTTP) were released; in 1995, the US Department of Commerce took over the Internet Assigned Numbers Authority from the Department of Defense, and the Internet began its commercialization process; in 1998, the Internet Corporation for Assigned Names and Numbers (ICANN) was established
Internet services	In 1994, Yahoo Inc. was founded and began providing search engine services, and Amazon was founded and began providing online book sales services; in 1998, Google Inc. was founded and began providing search engine services; in 2004, Facebook (later renamed Meta) was founded and began providing social network services

processing, display, and distribution of information. These technologies influence each other and iterate rapidly, having a huge impact on human society and bringing about the fourth information technology revolution.

Many new industries and categories have emerged around the processing of information. In 2006, the UN published the revised 4th edition of the "International Standard Industrial Classification of All Economic Activities", which newly established Category J "Information and Communications", including publishing activities (category 58), production of films, videos and television programs, and publishing of sound recordings and musical works (category 59), radio and television broadcasting (category 60), telecommunications (category 61), computer programming, consulting and related activities (category 62), information service activities (category 63), and other subcategories. In addition, in Category C "Manufacturing", the manufacturing of computers, electronic products and optical products (category 26) was established.

Table 1.7 Major technological advances in the field of mobile communications

Technologies	Representative events
1st-Generation Mobile Communications (1G)	In 1978, Bell Laboratories successfully developed the Advanced Mobile Phone System (AMPS), marking the official entry of the 1G system, which is mainly characterized by analog cellular networks, into the historical stage; the capacity of the 1G analog cellular system is very limited, supporting only voice calls and not providing data and roaming services
2nd-Generation Mobile Communication (2G)	In 1990, 2G systems characterized by digitalization began to be widely deployed around the world; 2G systems mainly include the Global System for Mobile Communications (GSM) proposed by Europe and the interim standard-95 (IS-95) system proposed by the US; 2G systems have seen significant improvements in performance and capacity, with improved voice quality and confidentiality, and have transmission rates approaching 1 Mbps (megabits per second)
3rd-Generation Mobile Communication (3G)	Around 2000, 3G systems began to be used commercially on a large scale; 3G systems are centered on data services and realize real mobile multimedia communications with transmission rates of about 3 Mbps; the time division-synchronous code division multiple access (TD-SCDMA) system proposed by China is listed as one of the three major international standards for 3G
4th-Generation Mobile Communication (4G)	Around 2010, 4G systems began to be put into commercial use; 4G systems have greatly improved transmission speeds and can meet the needs of applications such as games, high-definition mobile TV, and video conferencing, the maximum transmission rate can reach 100 Mbps; the time division-long term evolution-advanced (TD-LTE-Advanced) proposed by China is listed as one of the international standards for 4G

1.2 Features of ICT Industry

1.2.1 Scope of ICT Industry

China divides the industries of the entire society into three major sectors. The primary industry includes agriculture, planting, animal husbandry, fishery and other industries. The secondary industry includes mining, construction, machinery manufacturing and other industries. The tertiary industry includes education, health, finance, transportation and other service industries. Although the economic activities of the information industry have long appeared, they are still attached to other industries. For example, the information equipment manufacturing industry initially belonged to the secondary industry, and the consulting service industry initially belonged to the tertiary industry. With the widespread application of information technology in various fields of economy and society, the economic activity process of the information industry has gradually separated from the corresponding process of other industries and formed an independent industry. The US Department of Commerce

Table 1.8 Basic classification of ICT industry

Subclass names	Scopes
Basic telecommunications industry	It provides basic communications and transmission services for the public, including traditional telegraph, telephone and other services, wired broadband data communications services for the Internet, and mobile communications services for individuals
Information services industry	It includes traditional information services and new information services: traditional information services include publishing, postal services, document information, radio and television, etc.; new information services, also known as electronic information services, are information services that use electronic information technologies such as computers and modern communications as the main processing means, providing services such as information production and supply, information consultin, network operations and network value-added services, etc
Information development industry	It includes software industry, database development industry, electronic publishing (including games, animation, etc.), application program development and other information development businesses, etc
Information manufacturing industry	With electronic computers and communications equipment manufacturing as its main content, it includes the electronics industry (electronic components, devices, complete machine manufacturing), communications and network equipment manufacturing, other information equipment manufacturing, etc

gave the definition of the information industry in the Digital Economy 2000 in accordance with the Standard Industrial Classification (SIC) issued by the US in 1987: the information industry should be composed of hardware industry, software industry, service industry, communication equipment manufacturing industry and communication service industry. Table 1.8 lists the basic classification of ICT industry.

1.2.2 Position of ICT Industry

The ICT industry is a basic industry. Basic industries refer to industries that support and carry the national economy and social development. Their development scale and level restrict the development speed and quality of the entire national economy. A World Bank research report shows that the higher the level of ICT development of an economy, the faster its economic growth rate. Statistical laws show that for every 10% increase in the penetration rate of high-speed broadband Internet, the per capita GDP (gross domestic product) in developing economies will increase by 1.38%, and for every 10% increase in the penetration rate of mobile phones, the per capita GDP in developing economies will increase by 0.81%.

The ICT industry is a leading industry. Leading industries refer to industries that have a guiding role and a significant role in promoting effect in the development of other industries in the national economy. The input–output ratio of information technology in the transformation of traditional industries can reach more than 1:10 in developed countries. The combination of ICT with machinery, energy, transportation, construction, textiles, metallurgy and other industries has continuously spawned new industrial categories and broader technological fields. The ICT industry has spawned and driven the development of emerging industries, such as the digital content industries (e.g., online publishing, online games, digital music, and mobile media), and the new Internet service industries (e.g., e-commerce, modern logistics, software and service outsourcing).

The ICT industry is a strategic industry. It has become a strategic industry for countries to compete for technological, economic, talent and military dominance and commanding heights. In the field of ICT, the intensity of competition in technical standards is far greater than in other industries. If a country or a company can control the standards, it will have a competitive advantage that other countries or companies cannot surpass. Additionally, the manufacturing of electronic information products and the software industry are also strategic sectors vital to national security, serving as the fundamental guarantee for a country's network and information security.

1.2.3 Features of ICT Industry

The ICT industry is an intellectually intensive industry. It is an industry that collects, produces and manages information. It is characterized by the substantial development of knowledge and technology, with a focus on intellectual labor. It is also an industry with the most concentrated and fierce competition for international patent technology. The top three high-patent-intensive industries in the US are the computer industry, the communications industry, and the electronics industry. For example, in order to seize the source and entrance of mobile Internet and big data, Google has launched patent acquisitions around intelligent terminals, and it acquired 17,000 patents from Motorola for 12.5 billion USD.

The ICT industry is a highly innovative industry. It is based on scientific research and exhibits characteristics of high innovation and a rapid pace of updates. The innovation speed of the ICT industry is also unmatched by other industries: the number of information technology doubles every three years, more than 300,000 new patents are added annually, and the effective lifespan of research data averages only five years. The rapid development of information technology has resulted in a shortened product lifecycle for information products, leading to increasingly frequent product updates. As Bill Gates, the former president of Microsoft, reminded his employees: Microsoft is always 18 months away from bankruptcy.

The ICT industry is an industry characterized by high risk and high returns. Researching and developing information products require substantial capital, and due to the uncertainty of the success probability in creating inventions, substantial investments may end up yielding no returns. The research and development of information products entails high fixed costs and low marginal costs. However, when it reaches the stage of mass production with standardized scales, the investment costs tend to be relatively lower. Information products often benefit from economies of scale, allowing manufacturers to reap much higher profit margins than those of traditional industries once they achieve significant sales volumes.

1.2.4 Development Laws of ICT Industry

Compared with traditional industries, the ICT industry has its own development laws, development paths, and development models. Among development laws, the most famous and far-reaching ones are Moore's Law, Metcalfe's Law, and "smiling curve".

1.2.4.1 Moore's Law

Moore's Law was proposed by Gordon Moore, one of the founders of Intel. In 1965, Moore was invited to write a review report entitled "Cramming More Components onto Integrated Circuits" for the 35th anniversary special issue of the journal Electronics. When Moore plotted the data, he found that the number of transistors that each generation of chips could accommodate was about twice that of the previous generation of chips, and the time interval between two generations of chips was within 18 to 24 months. If chip capacity develops according to this trend, computing power will increase exponentially relative to the time period.

Moore's Law summarizes the law of the speed of information technology development. It points out that the performance of integrated circuit chips has indeed been greatly improved in the past 30 years. When looking forward to the future of integrated circuit chips, information technology experts believe that Moore's Law may still apply. However, as transistor circuits gradually approach the performance limit, Moore's Law will come to an end in this specified area.

1.2.4.2 Metcalfe's Law

Metcalfe's Law was proposed by Robert Metcalfe, the founder of the American network equipment company 3Com Corporation. Its basic content is that the value of a network is equal to the square of the number of network nodes, that is, the commercial value of a network is proportional to the square of the number of the connected users. Metcalfe's Law reveals that the commercial value of a network grows quadratically with the growth of the number of the connected users.

Metcalfe's Law states that when a technology has established a necessary user scale, its value will explode. Since the 1990s, the Internet has not only shown this extraordinary growth trend, but has also widely penetrated and expanded into various fields of the economy and society. As the number of Internet users increases, network resources grow exponentially. Currently, with the widespread application of mobile Internet and social media, Metcalfe's Law has been fully confirmed.

1.2.4.3 Smiling Curve

Mr. Shi Zhenrong, the founder of Acer Group, proposed the famous "smiling curve" theory (see Fig. 1.3) in 1992 in order to "rebuild Acer". The "smiling curve" divides the industrial chain of the manufacturing industry into three stages, namely research and development (R&D), manufacturing, and marketing. It believes that the added value of the industry is more reflected at the two ends, namely R&D and marketing, while the added value of the manufacturing stage in the middle is the lowest. Researches on the modern industrial value chain also show that the profit margins of enterprises at the two ends of the industrial chain are 20% to 25%, while the profit margins of processing and production enterprises in the middle are only about 5%. This form with the two ends facing up and the middle facing down is like a smile logo.

From a global perspective, developed countries and regions dominate the development trend and pattern of the ICT industry and are at the high end of the value chain. With its strong economic strength, large-scale basic research, and perfect venture capital mechanism, the US is committed to the research, development and application of information technology, constantly launching new technologies, and maintaining its leading position in the world. European countries such as the UK and Germany are also committed to R&D and standard setting in some areas of the ICT industry, and are at the high end of the value chain. South Korea, Singapore, and other countries and regions are in the middle of the ICT industry value chain based on production technology.

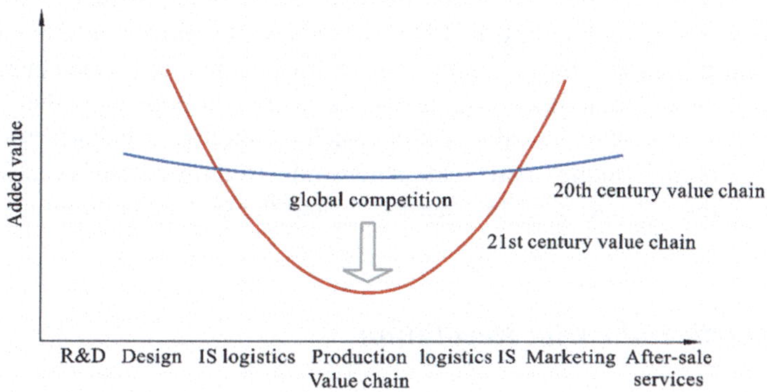

Fig. 1.3 Smiling curve. *Image Source* Shi Zhenrong, *Acer's Century of Change*, 2005

1.3 Revolution of Technology and Industry

1.3.1 New-Generation ICT

After 2010, ICT ushered in a new round of development boom. Artificial intelligence (AI), the Internet of Things (IoT), and the new generation of mobile communications technologies are considered to be another major round of changes in the ICT industry. These new technologies will subvert traditional communications and computing models, and drive the ICT industry to a higher level. These new technologies support each other, evolve in groups, accelerate breakthroughs, and have been widely used. The representative new-generation ICT are shown in Table 1.9.

The new generation of ICT is accelerating its horizontal spread to other economic sectors and interpenetrating and integrating with the physical world. The fundamental and leading role played by ICT in high-tech industries is becoming more and more significant.

First, the connotation of information space is continuously enriched and profound, and it has become an important carrier to realize the control of matter and energy. The deep integration of information and physics has profoundly changed the perception, analysis, decision-making and execution process in the traditional control loop, and the control system has taken a substantial step towards intelligence. Artificial intelligence technology can achieve autonomous cognition and learning, thereby optimizing and reconstructing the production and industrial operation mode, and creating greater incremental value.

Second, the systematic development of information technology has become an important form of technological competition in the field of electronic information. The technologies of key links such as perception, transmission, storage, and computing are constantly cross-integrating, and the systematic development model is becoming increasingly obvious. The Internet and big data have given rise to a new paradigm of data-driven engineering technology development, triggering profound changes in scientific research methods and organizational methods.

Third, as a general technology, ICT is combined with various fields and generates new technical directions. For example, information technology is accelerating the cross-integration with manufacturing, materials, energy, biology and other technologies, giving rise to new technical directions such as intelligent manufacturing, new materials, and bioinformatics, and creating new industrial forms and business models such as the Industrial Internet, "Internet Plus", and the sharing economy.

1.3.2 Fourth Industrial Revolution

Science and technology are the main driving force for social and economic development. Every rapid economic growth is closely linked to a large-scale industrial revolution. The industrial revolution is based on new scientific discoveries and marked

Table 1.9 The representative new-generation ICT

Fields	Content of technologies
5th-Generation Mobile Communications Technology (5G)	The 5th-Generation Mobile Communications Technology referred to as 5G, is a new generation of mobile communications technology with the characteristics of high speed, low latency and large connection, and the three major application scenarios of 5G defined by the International Telecommunication Union (ITU) are enhanced mobile broadband (eMBB), ultra-reliable and low-latency communication (uRLLC) and massive machine type communication (mMTC)
Industrial Internet	The Industrial Internet is a new type of infrastructure, application model and industrial ecology that deeply integrates the new-generation ICT with the industrial economy; through the comprehensive connection of people, machines, objects, systems, etc., it builds a new manufacturing and service system covering the entire industrial chain and the entire value chain, and its network system includes three parts: network interconnection, data interoperability, and identity resolution
IoT	The IoT is a network and its application system with various sensor devices as nodes; it can realize the identification of physical objects, the perception of the environment, the control of equipment, etc., and improve the level of automation in production and life; its representative technologies include radio frequency identification (RFID), sensor networks, machine-to-machine (M2M), etc
Cloud computing	Cloud computing is a model that provides computing resources (servers, databases, storage, etc.) and software services (platforms, architectures, applications, etc.) to users on demand through the Internet, and cloud computing service providers use distributed computing, virtualization and other technologies to achieve automated management, sharing and allocation of computing resources
Big data	Big data refers to the technology of processing, storing, analyzing and managing high-speed, large-scale and diverse data sets; with the huge amount of data accumulated by the Internet technology, big data technology needs to process a large amount of structured, semi-structured and unstructured data, which is often achieved by cloud computing
Blockchain	Blockchain is a distributed storage system based on the Internet; it uses technologies such as point-to-point transmission, consensus mechanism, and encryption algorithm, so it is anonymous and tamper-proof, and provides a secure and reliable decentralized storage solution

(continued)

Table 1.9 (continued)

Fields	Content of technologies
Metaverse	The metaverse is a virtual world system that is linked and created by scientific and technological means, and maps and interacts with the real world; it involves technologies such as virtual reality, augmented reality, digital twins, and human–computer interaction, and can form a digital living space with a new social system
AI	AI is a technology that uses computers to simulate certain human thinking processes and intelligent behaviors (such as learning, reasoning, thinking, planning, etc.), involving fields such as robotics, language recognition, image recognition, natural language processing, computer vision, machine learning and expert systems; in recent years, large AI models have adopted machine learning models with ultra-large-scale parameters (usually more than 1 billion) and ultra-powerful computing resources, which have excellent performance in natural language processing, image processing, etc

by the outbreak of technological revolution. New technologies are widely used in human production and life, thus triggering changes in industrial structure, economic situation, social division of labor system, global political and economic structure, and changing social and cultural values and the psychological state of social members.

The First Industrial Revolution was marked by the invention of the steam engine, led by the UK, and characterized by the replacement of human and animal power with mechanical power, which led to large-scale factory production and triggered the mechanization of the textile industry and the transformation of the metallurgical industry. The Second Industrial revolution was mainly marked by the invention and application of internal combustion engines and electric power technology, led by the US and Germany, and realized the electrification of production and life with electric power as the power, gave birth to the automobile manufacturing industry and petrochemical industry, and promoted the development of innovative industries such as railway transportation and shipbuilding. The Third Industrial Revolution was the information technology revolution, represented by the development and application of electronic computer technology, and continued to be led by the US, realizing the automation, informatization and modernization of production and life.

At present, a new technological revolution represented by AI, big data, the IoT, space technology, biotechnology, and quantum technology is underway. Ulrich Sendler, the author of *Industry 4.0: The Coming Fourth Industrial Revolution*, calls this technological revolution the Fourth Industrial Revolution. The impact of various new technologies adopted in this technological revolution on human beings is systematic and holistic. Various emerging technologies are highly integrated and mutually infiltrated to form a complete technological system, which has the characteristics of synchronization, systematicity, and integrity in the transformation of human society, and has created new products, new formats, and new models that cross the boundaries of traditional industries. The Fourth Industrial Revolution is an intelligent revolution,

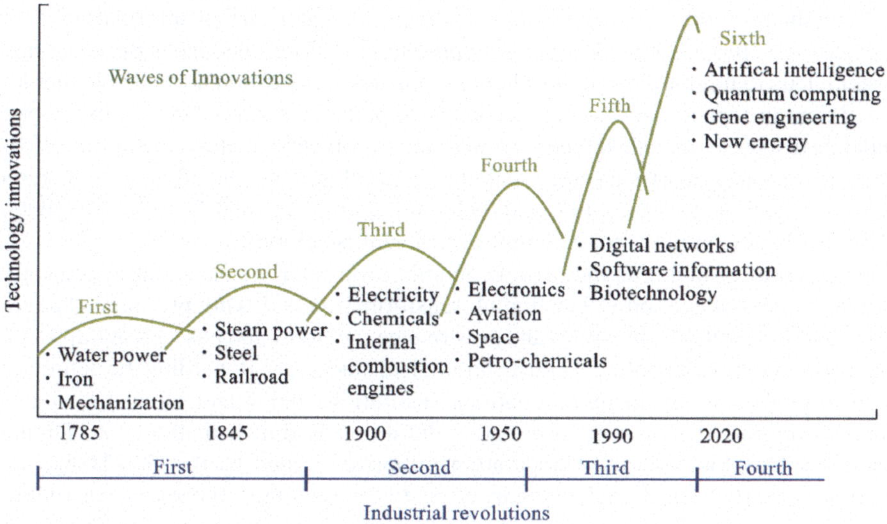

Fig. 1.4 Technology innovations and industrial revolutions

represented by genetic engineering, quantum computing, new material technology, new energy technology, virtual reality, etc., to achieve comprehensive intelligence of production and living systems, and bring about major changes in the development mode of the economy and society. For example, the Internet is gradually moving towards the IoT, traditional enterprises are also evolving into digital enterprises, traditional manufacturing is transforming into "intelligent manufacturing", and intelligent robots will appear in all areas of human production and life. Figure 1.4 shows the history of technology innovations and industrial revolutions.

1.4 New Digital Economy

1.4.1 Scope of Digital Economy

From the perspective of industrial revolution history, the digital economy can be traced back to the emergence of electronic computers in the 1950s. The early manifestation of the digital economy is the Internet economy. In 1994, the concept of "digital economy" appeared. The book *Digital Economy* by the famous economist Don Tapscott was also published in this year. This is the first book on digital economy, which predicts the new form of economic development in the Internet era, such as the emergence of e-commerce community. Nicholas Negroponte also described the social life scenes in the digital age in his book Being Digital.

With the promotion and application of the Internet and various information technologies, data has become a key factor of production fully embedded in the economic system. Due to its characteristics of being replicable, easy to spread, and able to transcend time and space, data has unprecedented application potential in various sectors, and data resources are called "the new oil of the twenty-first century". Data can effectively drive the network sharing, collaborative development, and efficient utilization of factors such as labor, capital, and technology, improve total factor productivity, and provide new impetus for economic and social development.

The digital economy is the main economic form after the agricultural economy and the industrial economy. The digital transformation is driving profound changes in production methods, lifestyles and governance methods, and has a great impact on the world's economic, political and technological landscape. Regarding the definition of the digital economy, many international institutions and organizations have made generalizations. The most representative definition is drawn in the "G20 Digital Economy Development and Cooperation Initiative" issued by the G20 Hangzhou Summit in 2016: "The digital economy refers to a broad range of economic activities that includes using digitized knowledge and information as the key factor of production, modern information networks as the important activity space, and the effective use of ICT as an important driver for efficiency-enhancing and economic structural optimization." The digital economy represents the sum of a series of economic activities of production, circulation and consumption around data, which serves as the key factor of production.

In the digital economy era, the process of gradually transforming traditional information industries into digital industries with digital as the main factor of production to promote the development of a new generation of information industries is called "digital industrialization", which refers to the industrialization, commercialization and marketization of data elements. The informatization and digital transformation of traditional industries is called "industrial digitization", which means using modern digital information technologies, like the Internet and AI to transform traditional industries in all perspectives, directions, and chains, so that digital technology can be deeply integrated into and develop with all walks of life in the real economy.

In May 2021, the National Bureau of Statistics of the People's Republic of China released the "Statistical Classification of Digital Economy and Its Core Industries (2021)", which for the first time defined the industrial scope of the digital economy: digital product manufacturing industry, digital product service industry, digital technology application industry, digital element-driven industry, and digital efficiency improvement industry. These five categories can be grouped into two aspects: digital industrialization and industrial digitization.

Digital industrialization: The first four categories above are the digital industrialization part, namely the core industries of the digital economy, which refer to the various economic activities that provide digital technologies, products, services, infrastructure and solutions for the digital development of industries, as well as all kinds of economic activities that are completely dependent on digital technologies and data elements. They correspond to the 26 major categories, 68 medium categories and 126 small categories in the "National Economic Industry Classification" and are the foundation for the development of the digital economy. The scope of digital industrialization is basically the same as that of the traditional ICT industry, including basic telecommunications industry, Internet industry, information service industry and electronic information manufacturing industry.

Industrial digitalization: The fifth major category of "digital efficiency improvement industry", namely the industrial digitalization part, refers to the application of digital technologies and data resources to traditional industries to increase output and improve efficiency. It is the integration of digital technologies and the real economy. This part covers digital application scenarios such as intelligent manufacturing, smart agriculture, and smart transportation, corresponding to 91 major categories, 431 medium categories, and 1256 small categories of the "National Economic Industry Classification". This reflects that digital technologies have been and will be further integrated into various industries of the national economy.

1.4.2 Global Digital Strategies

The role of data in promoting the economy is becoming increasingly prominent, and using ICT to promote the digital transformation of traditional industries has become a consensus among countries. In recent years, the value of data elements has been released in some countries and regions, such as the EU, the UK, the US, Brazil, Japan, etc. The value of data market has continued to increase, and relevant data indicators such as data professionals and data companies have been rapidly developed. In 2020, the value of the data economy in the 27 EU countries reached approximately 355.1 billion euros, an increase of 9.3% over 2019. In order to further promote the development of the ICT industry and the digital economy, more and more countries have taken various positive measures to protect and develop their own ICT industries. The Digital Economy Outlook 2020 published by the Organization for Economic Cooperation and Development (OECD) shows that, in the digital economic policy survey conducted in 37 countries, 34 countries had formulated national overall digital strategies.

In order to maintain its absolute global leadership in the field of electronics information, the US has continuously strengthened the government's guidance on the development direction of information technology and released relevant development plans or strategies in a timely manner. In 1998, the US Department of Commerce released the "Emerging Digital Economy" series of reports; in recent years, it has successively released the "Initiative on the Digital Economy", the "Federal Data Strategy 2020 Action Plan", and the US Innovation and Competition Act of 2021, repeatedly emphasizing the important role of the digital economy in the development of the national economy. In October 2020, the US released the "National Strategy for Critical and Emerging Technologies", redefining 20 key and emerging technologies, and proposing to fully maintain the global leadership of the US in cutting-edge technologies such as quantum physics and AI.

In order to rebuild Europe's international competitive advantage in the field of ICT, the EU has actively laid out the innovative development of information technology. The "Horizon 2020" scientific research plan was implemented from 2014 to 2020, strengthening the innovation and application of information technology, of which investment in the field of ICT accounts for 46% of the total investment. After the EU proposed the European data strategy in 2014, it successively launched the "Digitising European Industry Initiative", the European Artificial Intelligence Act, and other documents, and released the "Shaping Europe's Digital Future" in February 2020, as well as the "2030 Digital Compass: the European Way for the Digital Decade" in March 2021, which comprehensively planned the EU's digital development path from 2021 to 2030.

1.5 Summary

The level of information technology has an important impact on social and economic life. Papermaking and printing techniques in ancient China provided a carrier for the recording and communication of traditional information, and promoted the development and communication of ancient world civilizations. Benefiting from the development of modern science and technology, various technologies were invented based on the electronic devices and radio waves, including telegraphs, telephones, and televisions, bringing human society into an era of global communication and exchange. Over the past half century, information technologies such as integrated circuits, computers, the Internet, and mobile communications have developed rapidly, driving the rapid progress and development of human society and profoundly affecting People's work and life. The development history of ICT industry shows that ICT, as a general technology, has a strong driving role and radiation effect on the development of other industries and the society, and is an economic growth "multiplier", a development mode "converter", and an industrial upgrading "booster".

At present, a new generation of ICT, which is represented by cloud computing, big data, 5G, Industrial Internet, artificial intelligence, blockchain, etc. is developing rapidly, and is also promoting the Fourth Industrial Revolution on a global scale. The widespread application of ICT in all walks of life has promoted the arrival of a digital society. Data has become an important factor of production and strategic resource, and is a new driving force for optimizing resource allocation, promoting the continuous upgrading of traditional industries, and improving social labor productivity. Developing a digital economy with data as the core element has become a consensus around the world. Meanwhile, formulating corresponding ICT development plans and digital economic development strategies has also become a consensus among major industrialized countries.

Chapter 2
Overview of Digital China

2.1 Brief History

2.1.1 Self-Reliance: Initial Establishment of China's ICT Industry (1949–1978)

On October 1, 1949, Chinese President Mao Zedong announced the founding of the People's Republic of China at Tian'anmen Square in Beijing. China put an end to more than a century of suffering under oppression and aggression, finally gaining the opportunity for independent development. At that time, China had just emerged from decades of war, leaving the country war-torn and in a state of ruin, with everything needing to be rebuilt. It was a backward agricultural country with almost no industrial foundation, being unable to produce even steel gasoline barrels. Electronic products were mainly dependent on imports. Only a few major cities had telephones, with just 260,000 telephone users nationwide and a wired telephone penetration rate of only 0.05%. There were fewer than 50,000 scientific and technical personnel across the country, and very few experts in electronic technology.

China built an independent electronics industry from scratch. Starting from 1952, China established a number of electronic manufacturing enterprises through transformations and new constructions (see Table 2.1). Among them, Nanjing Radio Factory successfully developed China's first fully domestically produced electronic tube radio, the "Red Star" Model 502 in 1953, ending the history of relying on foreign components for radio assembly; Beijing Electron Tube Factory, with assistance from the Soviet Union, established an electron tube production line and became the largest electron tube factory in Asia during the 1960s. In 1963, the Fourth Ministry of Machine Industry (later renamed the Ministry of Electronics Industry in 1982) was founded to coordinate the national electronics industry, marking the establishment of China's information and communications technology (ICT) industry as an independent industrial sector.

W. Liu et al. (eds.), *Digital China*, Introduction to China's S&T Innovation,
https://doi.org/10.1007/978-981-95-6240-4_2

Table 2.1 Development history of electronic manufacturing in China (1950s–1970s)

Fields	Representative developments
Semiconductor	In 1952, Nanjing Electric Factory successfully developed China's first radio tube; In 1956, the Institute of Applied Physics of the Chinese Academy of Sciences developed China's first triode; In 1957, the Beijing Electron Tube Factory produced China's first germanium single crystal; In 1968, Shanghai Radio 14th Factory manufactured PMOS integrated circuits
Electronic product	In 1953, Nanjing Radio Factory successfully developed the first domestically produced electronic tube radio; In 1958, China's first black-and-white TV was born and named "Beijing"; In 1960, China independently developed the first 1000-door crossbar automatic telephone exchange equipment; In 1970, China's first artificial satellite "Dongfanghong-1" was successfully launched
Computer	In 1958, the Institute of Computing Technology of the Chinese Academy of Sciences successfully developed China's first small-scale general-purpose computer, the 103; In 1965, the Institute of Computing Technology of the Chinese Academy of Sciences successfully developed the first large-scale transistor computer, the 109B; In 1974, Tsinghua University and other units jointly developed the DJS-130 small multi-purpose computer using integrated circuits

China also built a scientific research team and talent cultivation system for the electronics industry from nothing. In November 1949, the Chinese Academy of Sciences (CAS) was established, forming a strategic scientific and technological backbone of China. In electronics-related fields, CAS successively established the Institute of Applied Physics (1950), Institute of Electronics (1956), Institute of Computing Technology (1959), and Institute of Semiconductors (1960), initially forming a national research team covering electronic information-related disciplines. In 1952, China began to restructure its higher education system to meet national economic development needs. Electronic and communications-related departments were newly established in comprehensive universities, such as the Department of Radio Engineering at Tsinghua University in 1952, and the Department of Radio Electronics at Peking University in 1958. Specialized engineering colleges in telecommunications were also founded, including Beijing Post and Telecommunications Institute (now Beijing University of Posts and Telecommunications) in 1955 and Chengdu Institute of Radio Engineering (now University of Electronic Science and Technology of China) in 1956.

Under the complex and ever-changing international situation, Chinahas long relied on self-reliance and hard work. Over the past three decades, it has built a scientific and development team to develop electronic technology, communications technology, and semiconductor technology from scratch, established an independent electronics industrial system, and developed the ability to produce basic electronic devices such as simple electron tubes and transistors, as well as the ability to develop electronic products and computers.

However, due to limited overall national strength at the time, China's telecommunications infrastructure remained weak. By 1978, the national telephone network had not yet deployed automatic exchanges, and the manual exchanges were still used in many counties and rural areas. Long-distance transmission mainly relied on wires and analog microwaves. The national telephone capacity was 3.59 million lines, with 2.14 million users, and the telephone penetration rate was only 0.38%. Less than 1 in every 200 people had a telephone, which was less than 1/10 of the world average level at the time.

2.1.2 Reform and Opening-Up: Rapid Development of China's ICT Industry (1978–2010)

In the late 1970s, significant changes occurred in the international landscape. In the 1980s, peace and development gradually became the themes of the era. At the end of 1978, China began to implement the national policy of "Reform and Opening-Up"[1] and shifted its focus to economic development, ushering in a period of rapid development for its ICT industry. Since 1986, China has carried out a strategic plan for the development of the postal and telecommunications industry and the electronics industry in the form of Five-Year plans. In 1993, China established a new group, named National Economic Informatization Joint Conference, to plan the countrywide informatization construction, especially the national information network infrastructure. In 1998, the State Council of China merged the Ministry of Posts and Telecommunications and the Ministry of Electronics Industry into the Ministry of Electronics Industry to reorganize it into the Ministry of Information Industry (MII). In 2006, China formulated the 15-year "National Informatization Development Strategy 2006–2020" to implement the guiding ideology of the new industrialization road of "driving industrialization with informatization and promoting informatization with industrialization". In 2008, the MII and the State Council Informatization Office were integrated and reorganized to form the Ministry of Industry and Information Technology (MIIT).

In all previous Five-Year plans, the electronics industry and the information industry were the focus of industrial development, and promoting the penetration of the national telephone network and data communications network was the focus of information infrastructure layout. With the rapid development of China's information industry and the continuous improvement of information infrastructure, the information industry has gradually become an important driving force for promoting

---

[1] At the end of 1978, China began to carry out the national policy of internal reform and opening-up to the outside world, transforming the planned economic system into the socialist market economic system, and actively participating in international economic cooperation and competition under economic globalization.

industrial upgrading and the overall economy. Table 2.2 lists the plans of China's information industry development from 1986 to 2010.

The ICT industry had gradually become a pillar industry of China's national economy. In 2010, the sales revenue of China's above-scale electronic information manufacturing industry reached 6394.5 billion yuan, double the amount compared to 2005 (3101 billion yuan),with a compound annual growth rate of more than 15% in the past five years; the export of electronic information products accounted for more than 30% of the country's foreign trade exports. Colorful TVs, microcomputers, mobile phones and other major output of fully-assembled products reached 120 million units, 250 million units and 1 billion units respectively, all accounting for more than 40% of the global total output. Industrial concentration has been significantly improved, and the competitiveness of key enterprises represented by the top 100 enterprises has been significantly enhanced. From 2005 to 2010, the main business revenue of the top 100 electronic information enterprises increased from 964.3 billion yuan to 1535.4 billion yuan, and the sales revenue of Huawei, Lenovo, Haier and other companies exceeded 100 billion yuan.

The innovation system with enterprises as the main body has gradually formed. Significant progress has been made in the large-scale production technologies of thin-film transistor liquid crystal display (TFT-LCD) and plasma display panels (PDP). A number of mid-to-high-end integrated circuit products, such as central processing unit (CPU) and mobile communications chip, have made breakthroughs. Some special technologies, such as 65-nm advanced processes and high-voltage processes, have achieved mass production. Three-dimensional packaging and other new packaging technologies have been developed and applied. A complete industrial chain system for time division-synchronous code division multiple access (TD-SCDMA) technology has been established, and technological breakthroughs have been achieved in the field of 40 Gbps ultra-large-capacity optical transmission systems. Digital TV terrestrial transmission technology and digital audio and video encoding and decoding technology have reached international advanced levels. Table 2.3 lists China's achievements in communication infrastructure from 1978 to 2010.

In terms of basic communications facilities, China has accelerated the construction of basic communications networks, covering optical fiber, mobile, and satellite communications, and established state-owned communications operators such as China Telecom, China Mobile, China Unicom, and China Satcom. In 2010, the basic telecommunications operators had the optical cable lines with a length of 9.962 million kilometers, 188 million Internet broadband access ports, and 620,000 3G network base stations. They had 1.6 Tbps (terabit per second) of total international communication business export bandwidth, with 7 land-submarine optical cables and 20 terrestrial cables. In 2010, the total number of telephone users in China reached 1.153 billion, with a penetration rate of 86.5 per 100 people; the number of Internet users reached 457 million, with a penetration rate of 34.3%. Besides the users of fixed-line telephone and mobile phones, the total number of Internet users and broadband access users in China jumped to the first place in the world in 2008.

Table 2.2 Plans of China's information industry development (1986–2010)

Planning period	Representative plans
During the 7th Five-Year Plan period (1986–1990)	In December 1986, the General Office of the State Council forwarded the "Report Outline on Doing a Good Job in the Promotion and Application of Computers in China" by the Electronic Revitalization Leading Group, and the outline proposed that the electronics industry should quickly shift to a track based on microelectronics technology and with computers and communications equipment as the main body; In March 1986, the "7th Five-Year Plan for National Economic and Social Development of the People's Republic of China" proposed to give priority to the development of transportation and communications
During the 8th Five-Year Plan period (1991–1995)	In April 1991, the "Ten Year Programme of the People's Republic of China for National Socio Economic Development and the Outline of the 8th Five-Year Plan" proposed to give prominence to the development of the electronics industry and make it a leading industry in promoting the modernization of the industrial structure and the entire national economy
During the 9th Five-Year Plan period (1996–2000)	In March 1996, the "9th Five-Year Plan for National Socio Economic Development of the People's Republic of China and the Programme for the Long Term Objectives by 2010" proposed that the postal and telecommunications industries should be developed with emphasis to form a unified national communications network system; In April 1997, the National Information Work Conference adopted the "Outline of National Information 9th Five-Year Plan and Long Term Goals for 2010", proposing to establish a national Internet information center and an Internet exchange center, and include the Internet in the national information infrastructure
During the 10th Five-Year Plan period (2001–2005)	In March 2001, the "Outline of the 10th Five-Year Plan for National Economic and Social Development of the People's Republic of China" proposed to drive industrialization with informatization, accelerate the promotion of informatization, and increase the proportion of the information industry in the national economy; In October 2002, the "10th Five-Year Plan for National Economic and Social Development Informatization Development Key Special Plan" proposed to promote the application of information technology, enhance the level of informatization, and strengthen the construction of modern information infrastructure

(continued)

Table 2.2 (continued)

Planning period	Representative plans
During the 11th Five-Year Plan period (2006–2010)	In March 2006, the "Outline of the 11th Five-Year Plan for National Economic and Social Development of the People's Republic of China" proposed to vigorously develop core industries such as integrated circuits, software and new components in the electronic information manufacturing industry, and focus on cultivating information industry clusters such as optoelectronic communications, wireless communications, high performance computing and network equipment; In May 2006, the "National Information Development Strategy 2006 2020" was released, further emphasizing "promoting industrialization with informatization and driving informatization with industrialization" and proposing strategic actions such as the key information technology autonomous innovation plan

In terms of the Internet, the number of Internet websites increased from 694,000 at the end of 2005 to 1.91 million in 2010, with a capacity of nearly 1800 TB. In 2010, the scale of Internet service market exceeded 200 billion yuan, forming a group of backbone enterprises with international influence. In 2010, the scale of China's online retail users reached 161 million, and the transaction value reached 513.1 billion yuan, accounting for 3.3% of the total retail sales of social consumer goods. The growth rate of online retail value was 5.7 times that of the retail sales of social consumer goods in the same period. New payment services, such as online payment, mobile payment, and telephone payment, developed rapidly, and the scale of third-party electronic payment increased rapidly, reaching 1.01 trillion yuan in 2010. The rapid development of modern logistics industry and online retail drove the rapid development of express delivery services. In 2010, the business volume of express delivery service enterprises above designated size in China reached 2.34 billion pieces, and the business revenue reached 57.46 billion yuan.

2.1.3 Towards a New Era: Leapfrog Advancement of China's ICT Industry (2011 to Present)

After 2010, the world once again set off a wave of accelerating informatization development. The rapid popularization of ICT technology represented by the Internet had a profound impact on international politics, economy, society, and culture. Various countries accelerated the development and application of information technology, and the integrated information network evolved towards broadband, convergence and ubiquity. ICT technology, products, content, networks and platforms had accelerated their integration and development.

In 2017, the report of the 19th National Congress of the Communist Party of China (hereinafter referred to as the 19th CPC National Congress) clearly proposed

Table 2.3 China's achievements in communication infrastructure (1978–2010)

Fields	Representative events
Optical fiber communications	In 1982, China built its first optical fiber communications system project, with a total length of 13.3 km, a speed of 8.448 Mbps, and the ability to transmit 120 telephone lines, ushering in a new era of digital communications in China; In 1998, China's "eight vertical and eight horizontal" grid-shaped optical cable backbone network was completed two years ahead of schedule, covering cities above provincial capitals and 70% of prefecture-level cities nationwide, the national long-distance optical cable reached 200 000 km, and China formed a long-distance backbone communications network with optical cables as the main medium and satellites and digital microwaves as the auxiliary components; In 2006, the first terabit-class, 10 Gbps submarine optical cable system, the Trans-Pacific Direct Cable System, between China and the United States was completed
Mobile communications	In 1993, China's first GSM digital mobile phone communications network (2G) was launched; In 2001, the number of mobile communications users in China exceeded 120 million, surpassing the United States to become the largest country in terms of mobile phone users; In 2003, China had 255 million fixed-line telephone users and 257 million mobile phone users, with mobile phone users exceeding fixed-line telephone users for the first time; In 2009, China issued three 3G licenses (CDMA2000, WCDMA, and TD-SCDMA)
Satellite communications	In 1984, China's first geostationary orbit experimental communications satellite, the Dongfanghong-2, was successfully launched, opening the history of satellite communications in China; In 1986, China's domestic satellite communications network was officially completed, including five communications satellites; In 1997, China successfully launched the Dongfanghong-3 satellite, which was equipped with 24 C-band transponders and reached the advanced level of similar communications satellites internationally; In 2000, China successfully launched the first two geostationary orbit satellites of the BeiDou Navigation Satellite System-1 (BDS-1), providing positioning services for China
Internet	In 1994, the national computing and networking facility of China was officially connected to the Internet through a 64 Kbps (kilobits per second) international dedicated line, ushering in China's Internet era; In 1997, China included the Internet in its national information infrastructure construction plan, gradually built four backbone networks with international export capabilities, and established the operation and management system of China's national top-level domain name (.CN); In 2008, the number of Chinese Internet users reached 298 million, and the Internet penetration rate exceeded the global average

the development strategy of building a "Cyber Power", a "Digital China", and a "Smart Society". Since the 12th Five-Year Plan, China has gradually formed a wide-ranging and multi-layered planning system for national informatization. In addition to the regular plans established every five years, it has proposed different plans specifically for national information infrastructure construction, economic transformation and upgrading, and new technology development. The typical development plans of China's ICT industry from 2011 to 2020 are shown in Table 2.4.

The scale of the digital economy grows rapidly. From 2017 to 2021, the scale of China's digital economy increased from 27.2 trillion yuan to 45.5 trillion yuan, ranking second in the world, with a compound annual growth rate of 13.7%, and its share of GDP increased from 32.9% to 39.8%, becoming one of the main engines driving economic growth (see Fig. 2.1).

The scale of digital industry grows rapidly. From 2017 to 2021, the revenue of above-scale computer, communications and other electronic equipment manufacturing industries increased from 10.6 trillion yuan to 14.1 trillion yuan; the revenue of large-scale enterprises in software industry increased from 5.5 trillion yuan to 9.5 trillion yuan; the revenue of above-scale enterprises in Internet and related service industries increased from 0.7 trillion yuan to 1.6 trillion yuan (see Fig. 2.2). From 2015 to 2020, the transaction volume of China's e-commerce increased from 21.8 trillion yuan to 37.2 trillion yuan. From 2015 to 2020, the information consumption of China experienced rapid growth, and the scale of China's information consumption increased from 3.4 trillion yuan to 5.8 trillion yuan. In 2016, the scale of the big data industry was 0.34 trillion yuan, and by 2020 it had exceeded 1 trillion yuan.

The network infrastructure is world-leading. By 2020, all prefecture-level cities in China had been built into optical network cities, and the access bandwidth of optical fiber networks in China had achieved exponential growth from 10 to 100 Mbps and then to 1 Gbps. The technical level and scale of China's mobile network had achieved a leap from "breakthrough in 3G" to "synchronization in 4G" and then to "leading in 5G" in the world. By the end of August 2022, the total number of 5G base stations in China had reached 2.102 million, and all prefecture-level cities and county towns had achieved full 5G coverage. The realization of "broadband access to every administrative village" had contributed to the completion of the historic goal of poverty alleviation[2] in China. The network infrastructure fully supported the Internet Protocol version 6 (IPv6). The number of national Internet backbone direct connection points had increased to 19, and the scale of data centers had exceeded 5.2 million standard racks. The BDS-3 had been launched, with a global positioning accuracy of less than 10 m. China's achievements in network infrastructure construction from 2011 to the present are shown in Table 2.5.

The number of Chinese Internet users increased from 688 million at the end of 2015 to 989 million at the end of 2020, the Internet penetration rate increased

[2] The "poverty alleviation" campaign is a large-scale and systematic strategic initiative implemented by China to improve the living conditions of poverty-stricken areas and impoverished people, and to promote economic development and social progress. On February 25, 2021, General Secretary of the CPC Central Committee Xi Jinping announced that China had accomplished the arduous task of eradicating absolute poverty. This is a great achievement in human history.

Table 2.4 Typical development plans for China's ICT industry (2011–2020)

Planning period	Representative plans
During the 12th Five-Year period (2011–2015)	In March 2011, "the Outline of the 12th Five-Year Plan for National Economic and Social Development of the People's Republic of China" was released, proposing to cultivate and develop strategic emerging industries such as the "new generation of information technology industry"; In August 2013, the State Council issued the "'Broadband China' Strategy and Implementation Plan", proposing that by 2020, the fixed broadband household penetration rate will reach 70%, the proportion of administrative villages with broadband access will exceed 98%, and the broadband access capacity of urban and rural households will reach 50 Mbps and 12 Mbps, respectively; In March 2015, the "Guiding Opinions of the State Council on Actively Promoting the 'Internet Plus' Action" was issued, proposing eleven special actions to promote "Internet Plus" innovation and entrepreneurship, collaborative manufacturing, modern agriculture, smart energy, inclusive finance, public services, efficient logistics, e-commerce, smart transportation, green ecology, and artificial intelligence (AI) to support China's economic transformation
During the 13th Five-Year Plan period (2016–2020)	In March 2016, the "Outline of the 13th Five-Year Plan for National Economic and Social Development of the People's Republic of China" was released, which once again emphasised the support for the development of strategic emerging industries, proposed to expand the space of the network economy, develop a modern Internet industry system, implement the national big data strategy, and strengthen information security protection; In July 2016, the General Office of the Communist Party of China Central Committee and the General Office of the State Council issued the "Outline of the National Information Development Strategy", proposing to build a "Cyber Power", and by 2025, the household penetration rate of fixed broadband will be close to the international advanced level, an internationally leading mobile communications network will be built, and seamless broadband network coverage will be achieved; In March 2017, the State Council issued the "New-Generation Artificial Intelligence Development Plan", pointing out that the artificial intelligence industry has become a new important economic growth point, and proposed to making China the world's major AI innovation center by 2030, and that the intelligent economy and intelligent society will achieve significant results

from 50.3% to 70.4%, and the number of mobile phone users increased from 1.11 billion to 1.643 billion, with the number of the users of 5G network growing from 0 to 355 million. Efforts to increase broadband speed and reduce fees have been continuously strengthened. In 2020, the average tariff level of fixed broadband and mobile data traffic dropped by more than 95% compared with 2015, and the average network speed increased by more than 7 times. The household penetration rate of fixed broadband increased from 52.6% at the end of 2015 to 96% at the end of 2020,

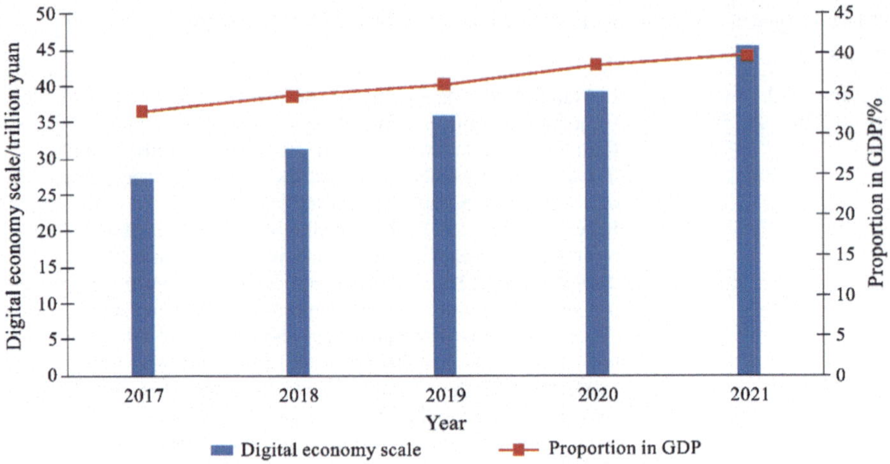

Fig. 2.1 The scale of China's digital economy and its proportion in GDP (2017–2021). *Data Source* China Academy of Information and Communications Technology

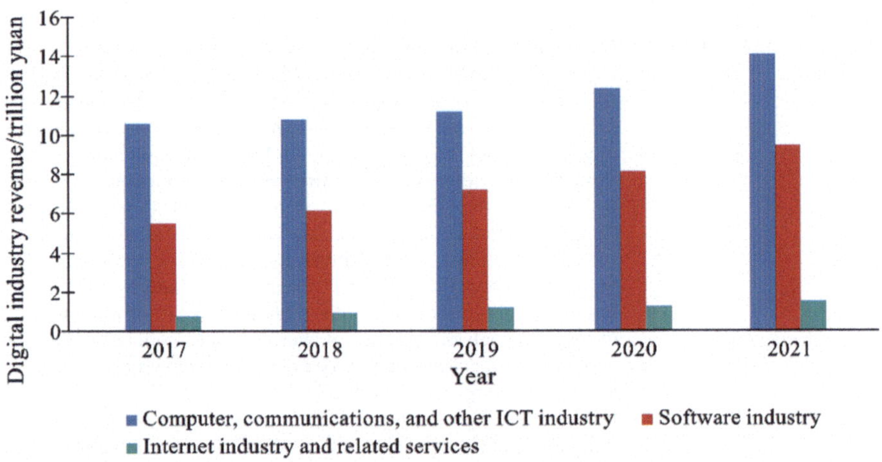

Fig. 2.2 Revenue growth of China's digital industry (2017–2021). *Data Source* Ministry of Industry and Information Technology of the People's Republic of China

the penetration rate of mobile broadband users increased from 57.4% at the end of 2015 to 108% at the end of 2020, and the proportion of administrative villages with optical fiber and 4G access had exceeded 98% nationwide. From the perspective of the proportion of broadband cost expenditure to per capita GDP, China's fixed broadband price index had risen from 71st in the world in 2017 to third in 2021, and the mobile broadband price index was lower than the global average for five consecutive years (see Fig. 2.3).

Table 2.5 China's achievements in network infrastructure construction (2011 to present)

Fields	Representative events
Optical fiber communications	In 2015, the number of fixed broadband users exceeded 200 million, ranking first in the world in terms of broadband user scale; In 2021, optical fiber broadband gigabit optical networks covered more than 120 million households, and the end-to-end broadband user experience speed reached 51.2 Mbps, ranking 18th among 176 countries and regions in the world
Mobile communications	In 2013, 4G commercial licenses were issued, ushering in China's 4G era; In 2015, the number of 4G users exceeded 380 million, and the scale of mobile broadband users (3G/4G) accounted for 60.1% of the total number of mobile users in China; In 2019, 5G commercial licenses were issued, ushering in China's 5G era
Satellite communications	In 2012, the BDS-2 was completed, including 14 satellites (5 geostationary orbit satellites, 5 inclined geosynchronous orbit satellites, and 4 medium earth orbit satellites), providing positioning services for the Asia–Pacific region; In 2017, ChinaSat 16 was launched, using a Ka-band multi-beam broadband communications system with a total capacity of 20 Gbps (Gigabit per second); In 2020, the BDS-3 was completed, with a total of 55 satellites, providing global positioning services
Internet	In 2012, mobile phones surpassed desktop computers to become the preferred terminal for Chinese netizens to access the Internet; In 2015, the number of Internet users reached 688 million, 90.1% of whom accessed the Internet via mobile phones, and the number of mobile Internet users reached 620 million; In 2015, online retail value ranked first in the world, reaching 3.88 trillion yuan; In 2020, e-commerce transaction value reached 37.2 trillion yuan

2.2 Current Scale and Level

After more than 40 years of reform and opening-up, China has made great achievements in the traditional economic sectors. From 2012 to 2021, the added value of the manufacturing industry increased from 16.98 trillion yuan to 31.4 trillion yuan, and its share of the world's total increased from 22.5% to nearly 30%, continuing to maintain China's position as the world's largest manufacturing country. From 2012 to 2021, calculated at constant prices, China's total industrial added value grew by an average of 6.3% per year, far higher than the global average annual growth rate of about 2% in the same period. Among the 500 major industrial products, China ranked first in the world in the output of over 40% of them. As to some important electronic products, such as personal computers, air conditioners, solar panels, and mobile phones, China's output accounted for more than half of the world's output.

In terms of digital economy, China also occupied a prominent position in the world, and the growth of China's digital economy was significantly higher than that of other major economies. In 2020, the digital economy of the United States (US),

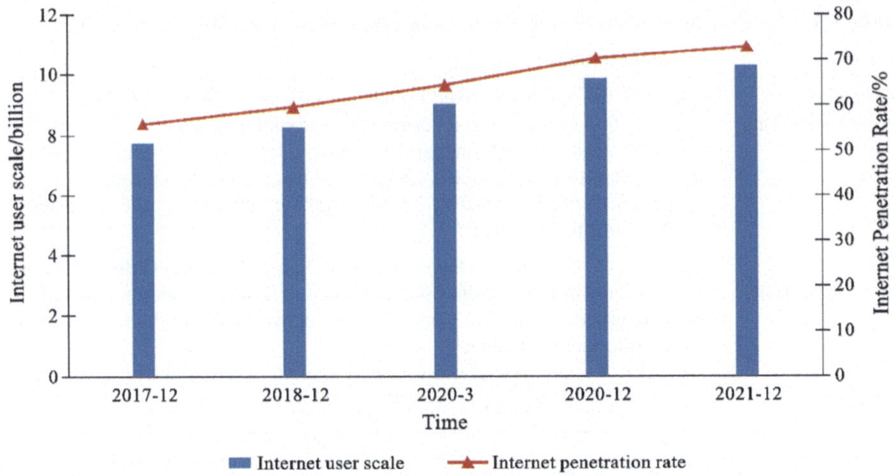

Fig. 2.3 The changing trend of China's Internet user scale and Internet penetration rate (2017–2021). *Data Source* China Internet Network Information Center

China, Germany, Japan, and the United Kingdom (UK) accounted for 79% of the world's total; the US digital economy ranked first in the world, with a scale of 13.6 trillion USD, accounting for 41.7% of the world; China's digital economy ranked second in the world, with a scale of 5.4 trillion USD; Germany, Japan, and the UK ranked third to fifth, and the scales were 2.54 trillion USD, 2.48 trillion USD, and 1.79 trillion USD respectively. In 2020, the added value of China's core digital economy industries accounted for 7.8% of GDP. The "Global Digital Economy Competitiveness Development Report (2020)" released by the Shanghai Academy of Social Sciences showed that the competitiveness of China's digital economy ranked third in the world, following the US and Singapore, and the gap to the US was shrinking year by year. Among the various sub-indicators, China's score in the digital industry ranked first in the world, and its leading advantage was expanding year by year. The evaluation results and ranking of global digital economy national competitiveness in 2020 are shown in Table 2.6.

In terms of digital infrastructure, China has built the world's largest, most widespread and technologically advanced information infrastructure. In terms of network facilities, mobile communications networks and optical fiber networks are the world's largest, with extensive coverage and leading technologies. In terms of computing power facilities, the scale of Internet data centers continues to grow rapidly, and a number of high-performance public computing platforms and intelligent computing centers dedicated to artificial intelligence are gradually taking shape. In terms of perception facilities, a pattern of coordinated development of narrowband Internet of Things (NB-IoT), 4G, and 5G networks has been formed, and network coverage capabilities continue to improve. The scale and level of China's information infrastructure in recent years are shown in Table 2.7.

Table 2.6 Evaluation results and ranking of global digital economy competitiveness in 2020

Ranking	Nations	Digital industry	Digital innovation	Digital facility	Digital governance	Total score
1	The US	46.76	80.18	69.89	86.54	70.84
2	Singapore	27.55	82.18	50.53	67.43	56.92
3	China	65.31	51.52	46.07	49.65	53.14
4	South Korea	12.85	68.48	46.33	65.4	48.27
5	The UK	20.32	65.37	33.42	72.8	47.98
6	Japan	12.66	73.45	39.09	63.4	47.15
7	Finland	3.07	85.54	33.51	63.77	46.47
8	Sweden	9.32	69.71	38.18	62.82	45.01

China's e-commerce ranked first in the world in terms of online retail volume and number of online shoppers. From 2013 to 2022, the scale of China's online retail market ranked first in the world for nine consecutive years consecutive. Calculated by the proportion of e-commerce sales in the entire national retail market, China reached 55.6% in 2022, ranking first in the world, while South Korea ranked second at 31.6%, the UK ranked third at 28.5%, the US was only 16.3%, and the average of Western European countries was 14.8% (see Fig. 2.4). China's online retail volume increased from 1.31 trillion yuan in 2012 to 13.1 trillion yuan in 2021, with an average annual growth rate of 29.15%; the e-commerce transaction value increased from 8 trillion yuan in 2012 to 42.3 trillion yuan in 2021, with an average compound annual growth rate of 20.3%. Relying on online e-commerce and mobile payments, the market scale of online car-hailing, online food delivery, digital culture, smart tourism, and other industries continually expanded.

2.3 Macro Strategy and Industrial Planning

2.3.1 National Development Strategy

The CPC is the core force leading China's reform and development. The overall strategy formed at the important meetings of the CPC will be reviewed by the National People's Congress and implemented by the governments at all levels. The 19th CPC National Congress (2017) and the 20th CPC National Congress (2022) formed the overall national informatization strategy, including the "Cyber Power" strategy for the development of the information industry, the "Smart Society" strategy for People's livelihood services and social governance, and the "Digital China" strategy for the overall digitalization of the national economy and society.

Table 2.7 The scale and level of China's information infrastructure in recent years

Types of infrastructure	Indicator	Scale and level
Network facilities	Broadband fixed network coverage level	All prefecture-level cities have been fully built into optical network cities, and by the end of October 2022, users with access bandwidth of 100 Mbps and above accounted for 93.8%, and users with access bandwidth of 1000 Mbps exceeded 80 million households
	Mobile communications network coverage level	The world's largest 5G network has been built, and by the end of September 2022, China had 2.22 million 5G base stations, accounting for more than half of the world's total, and 5G mobile phone users reached 510 million
	Satellite positioning and navigation service coverage level	In July 2020, the BDS-3 was officially launched, and one year later, high-precision services were launched in more than 20 countries, with a total number of users exceeding 2 billion
	IPv6 address resource scale	The network infrastructure has been fully evolved and upgraded to IPv6, and as of June 2022, the total amount of IPv6 address resources ranked first in the world, and the number of active IPv6 users reached 683 million
	Internet access level of residential users	As of June 2022, the number of Internet users in China reached 1.051 billion, with an Internet penetration rate of 74.4%, especially in rural areas, where the Internet penetration rate increased to 58.8%
	Internet access level in primary and secondary schools	By the end of 2021, the Internet access rate of primary and secondary schools reached 100%, the proportion of schools with export bandwidth of more than 100 Mbps reached 99.95%, and more than 210,000 schools had access to wireless networks
Computing facilities	Data center server scale	By the end of 2020, China had about 4.4 million standard data center racks in use, and the scale of cloud computing was close to 43 million cores, with an average shelf-load rate of about 50%, and a storage capacity of 800 EB (exabyte)
	Data center computing power level	By the end of 2021, China's total computing power scale exceeded 140 EFLOPS, with an average annual growth rate of more than 30% in the past five years, ranking second in the world in terms of computing power scale

(continued)

Table 2.7 (continued)

Types of infrastructure	Indicator	Scale and level
Perception facilities	NB-IoT coverage scale	China has built the world's largest NB-IoT, and deployed millions of NB-IoT base stations, basically achieving continuous coverage above the county level; as of December 2022, the number of mobile Internet of Things (IoT) connections reached 1.845 billion, and the scale of cellular IoT terminal users was rapidly approaching that of mobile phone users
	Number of IoT applications	By the end of 2022, four 10 million-level applications had been developed, including water meters, gas meters, smoke sensors, and tracking, and seven million-level applications had been developed, including white goods, street lights, parking, and agriculture

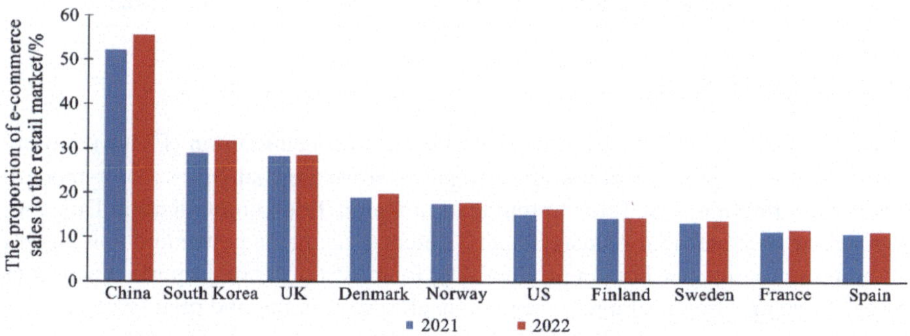

Fig. 2.4 The proportion of e-commerce sales in the retail market for various countries (2021–2022). *Data Source*: e-Marketer

2.3.1.1 Cyber Power

In February 2014, Xi Jinping, General Secretary of the CPC Central Committee, proposed to proceed from the international and domestic trends, make an overall layout, coordinate all parties, innovate and develop, and strive to build China into a "Cyber Power". In 2017, the report of the 19th CPC National Congress clearly proposed building a "Cyber Power". In 2022, the report of the 20th CPC National Congress further proposed to build a "Cyber Power". The strategic deployment of building a "Cyber Power" should be advanced in synchronization with the "Two

Centenary Goals".[3] Efforts should be following directions: the basic popularization of network infrastructure, significant enhancement of independent innovation capabilities, comprehensive development of the information economy, strong cybersecurity guarantees, and balanced network offensive and defensive capabilities. Main tasks of building a cyberpower include:

1. Reaching a world-leading level in Internet-related information infrastructure. The network scale and broadband penetration rate, the competitiveness of the network-related information industry, and cybersecurity capabilities will be at the world's leading level, and the construction and application of cloud computing, mobile Internet, big data, and the IoT will also be at the world's leading level.
2. Achieving independent control in key technologies related to the Internet. Fundamentally changing the situation where key technologies will be controlled by others requires major breakthroughs in key technology areas such as chip, operating system, and CPU.
3. Reaching a world-leading level in Internet applications. The Industrial Internet and the consumer Internet will be at a relatively high level as a whole, and e-commerce and e-government will be widely popularized and applied.

2.3.1.2 Smart Society

"Smart Society" was originally proposed by the European Union (EU) to explore how to use contemporary technological trends to solve the challenges facing modern society. The International Telecommunication Union Telecommunication Development Sector proposed that "Smart Society" refers to using the power and potential of technologies to improve human productivity, focus resources on important activities and relationships, and ultimately improve health, well-being and quality of life.

In 2017, the report of the 19th CPC National Congress clearly proposed to build a "Smart Society". It aimed to ensure People's livelihood, fully relying on information and intelligent means to provide the people with diversified, inclusive and equal public services. The main tasks of building a smart society include:

1. Intelligent infrastructure. Carry out intelligent transformation of public social infrastructure that involves the lives of the broad masses of the people, including transportation, power grids, water services, and logistics.
2. Informatization of planning and management. Through the comprehensive application of technologies such as City Information Model (CIM) and Geographic Information System (GIS), the digitization and precision of urban and rural infrastructure management such as urban and rural planning, land use, urban and rural pipe network, landscaping, and environmental protection will be promoted.

---

[3] The "Two Centenary Goals" are, to build a moderately prosperous society in all respects by the 100th anniversary of the founding of the Communist Party of China (2021), and to build China into a great modern socialist country that is prosperous, strong, democratic, culturally advanced, harmonious, and beautiful by the 100th anniversary of the founding of the People's Republic of China (2049).

3. Universalization of public services. Promote the informatization and intelligence of medical care, education, culture, tourism and other fields, promote the sharing and flow of high-quality medical and educational resources, and promote the digitization of public welfare venues such as libraries and museums.

2.3.1.3 Digital China

With the continuous acceleration of the digitalization process in various fields of economy and society, data elements will play a greater role in economic operation efficiency and total factor productivity. All countries in the world regard digitalization as the focus of economic development, and adopt policies and increase investment.

In 2017, the report of the 19th CPC National Congress clearly proposed to build "Digital China". In 2022, the report of the 20th CPC National Congress further proposed to accelerate the development of the digital economy, promote the deep integration of the digital economy and the real economy, and create a digital industry cluster with international competitiveness. The main tasks of building "Digital China" include:

1. Create new advantages in the digital economy. The first is to strengthen the innovative application of key digital technologies. Accelerate the advancement of research and development (R&D) breakthroughs and iterative applications in the fields of high-end chips, operating systems, key algorithms for AI, sensors, and general-purpose processors. The second is to accelerate the promotion of digital industrialization. Cultivate and expand emerging digital industries, and improve the industrial level of communication equipment, core electronic components, and key software. The third is to promote industrial digital transformation. Implement the action of "Using Data to Empower Wisdom on the Cloud" to promote the coordinated transformation of the entire industrial chain empowered by data. Deeply promote the digital transformation of the service industry, and cultivate new growth points such as crowdsourcing design, smart logistics, and new retail.
2. Accelerate the new pace of digital society construction. The first is to provide smart and convenient public services. Promote the digitalization of resources in public service institutions such as schools, hospitals, and nursing homes, and increase the intensity of resource sharing and sharing services. Actively develop online classrooms, Internet hospitals, smart libraries, etc. The second is to build smart cities and digital villages. Promote the construction of new smart cities by class and classification, incorporate IoT sensing facilities and communication systems into the unified planning and construction of public infrastructure, promote the intelligent transformation of municipal public facilities, and promote the construction of urban data brains. The third is to build a new picture of a better digital life. Promote the construction of smart communities, and develop online and offline integrated community life services and community governance.
3. Create a new ecosystem for digital development. The first is to establish and improve market rules for data elements. Cultivate standardized data trading platforms and market entities, promote basic legislation in the fields of data security

and personal information protection, and promote the safe and orderly flow of data across borders. The second is to create a standardized and orderly policy environment. Improve the management norms of the sharing economy, platform economy and new individual economy, and improve the statistical monitoring system of the digital economy. The third is to strengthen cybersecurity protection. Establish and improve the key information infrastructure protection system, and strengthen cybersecurity risk assessment and review. The fourth is to promote the construction of a community with a shared future in cyberspace. Promote the formulation of international rules for digital and cyberspace with the United Nations as the main channel and the United Nations Charter as the basis, and promote the establishment of a multilateral, democratic, and transparent global Internet governance system.

2.3.2 Government Planning Hierarchy

At present, China has established a planning system for the ICT industry and national informatization, covering all levels of government departments and all industries (see Table 2.8). Among them, the national plans mainly include:

1. Medium- and long-term development plans. These plans of about 15 years were made for the long-term development of national informatization. They could be the macro overall planning, such as the "National Informatization Development Strategy 2006–2020" issued in 2006 and the "Outline of the National Informatization Development Strategy" issued in 2016. They could also be the special plans for certain fields, such as big data, AI, and 5G.
2. Five-year plans. These plans were designed based on the overall five-year plan for national economic and social development. They specified the requirements for ICT development from various industry sectors, covering strategic emerging industries, electronic information manufacturing industries, digital transformation of traditional industries, social informatization and other fields. The competent departments of various industries would then formulate five-year information technology sub-item plans for each field based on the five-year plans.
3. Short-term plans. These plans were designed by the relevant national ministries and commissions in response to the development needs of a certain field. They lasted about three years, and were designed to organize and coordinate major national projects and key projects, and make special adjustments and improvements to related supporting policies, measures, and environment.

China had formed a complete planning system of "research-planning-execution-evaluation". In the process of developing plans, the central government and ministries will seek the opinions of relevant industry organizations and experts to ensure the scientific nature and feasibility of the plans. Major national plans, such as the five-year plans for national economic and social development, need to be reviewed by the National People's Congress and other legislative bodies. After the national plan

Table 2.8 Relevant plans for ICT industry and informatization during the 14th Five-Year Plan period

Planning types	Sub-item types	Issuing authorities	Phase
Medium-and long-term plans	Macro plans	General Office of the CPC Central Committee, Office of the State Council	Outline of the National Informatizationn Development Strategy
			Outline of the 14th Five-Year Plan for National Economic and Social Development and the Long- Range Objectives Through the Year 2035 of the People's Republic of China
	Special plans	State Council	Next-Generation Artificial Intelligence Development Plan
Five-Year plans	Macro-specific plans	Central Cyberspace Affairs Commission	14th Five-Year Plan for National Informatization
		State Council	14th Five-Year Plan for the Development of Digital Economy
		Office of the State Council	14th Five-Year Plan for the Modern Logistics Development
	Information industry plans	Ministry of Industry and Information Technology	14th Five-Year Plan for Information and Communications Industry Development 14th Five-Year Plan for the Development of the Big Data Industry 14th Five-Year Plan for the Development of Software and Information Technology Services
	Industry information plans	Ministry of Industry and Information Technology, etc	14th Five-Year Plan For the Development of Intelligent Manufacturing 14th Five-Year Plan for the Deep Integration Development of Informatization and Industrialization
		Ministry of Agriculture and Rural Affairs	14th Five-Year Plan for National Agricultural and Rural Informatization Development
		National Radio and Television Administration	14th Five-Year Plan for the Development of Radio, Television and Online Audiovisual
		Ministry of Commerce, etc	14th Five-Year Plan for E-commerce Development

(continued)

Table 2.8 (continued)

Planning types	Sub-item types	Issuing authorities	Phase
		Ministry of Transportation	14th Five-Year Plan for Digital Transportation
		Ministry of Housing and Urban–Rural Development, etc	14th Five-Year Plan for National Urban Infrastructure Construction
		Ministry of Civil Affairs	14th Five-Year Plan for Development of Civil Affairs Informatization
		National Health Commission, etc	14th Five-Year National Health Informatization Plan
		National Development and Reform Commission	14th Five-Year Plan for Promoting National Government Informatization
Short-term plans	Industry information plans	Ministry of Industry and Information Technology, etc	Three-Year Action Plan for the Construction of New Infrastructure for the Internet of Things (2021–2023) "Dual Gigabit" Network Collaborative Development Action Plan (2021–2023) Three-Year Action Plan for the Development of New Data Center (2021–2023) Action Plan for Green and Low-Carbon Development of the Information and Communication Industry (2022–2025) Action Plan for the Development of Basic Electronic Components Industry (2021–2023)
		Industrial Internet Special Working Group	Industrial Internet Innovation and Development Action Plan (2021–2023)
	Information application plans	Ministry of Industry and Information Technology, etc	Action Plan for the Integrated Development of Virtual Reality and Industry Applications (2022–2025) 5G Application "Sailing" Action Plan (2021–2023)
		Central Cyberspace Affairs Commission, etc	Digital Rural Development Action Plan (2022–2025)

is released, governments at all levels will develop corresponding plans at their level based on their spirit and in light of local realities, and submit them to the People's Congress at their level for deliberation and approval. Governments at all levels will list the goals in the plan as part of their annual work reports and summaries, report the progress of implementation to the People's Congress at the relevant level, and accept supervision.

China's hierarchical planning system can ensure that the central government's policies and plans can be effectively implemented throughout the country. As shown in Table 2.9, taking Hubei Province and Wuhan City as examples, after the State Council formulated the national five-year plan, Hubei Province and Wuhan City also formulated the relevant plans for their respective governments. After the national ministries and commissions formulated the five-year plan for one specified industry (such as the big data industry), the relevant industry authorities in Hubei Province further formulated the five-year plan for the industry in the province.

Table 2.9 Decomposition of China's information industry planning (taking Wuhan, Hubei as an example)

Planning types	Issuing authorities	Layer	Planning	Release date
Decomposition of macro development plan		National	Outline of the 14th Five-Year Plan for National Economic and Social Development and the Long-Range Objectives Through the Year 2035 of the People's Republic of China	March 2021
		Provincial	Outline of the 14th Five-Year Plan for National Economic and Social Development and the Long-Range Objectives Through the Year 2035 of Hubei Province	April 2021
		Municipal	Outline of the 14th Five-Year Plan for National Economic and Social Development and the Long-Range Objectives Through the Year 2035 of Wuhan City	April 2021
Decomposition of industry development plan	Ministry of Industry and Information Technology	National	14th Five-Year Plan for the Development of the Big Data Industry	October 2021
	Hubei Provincial Department of Economy and Information Technology	Provincial	14th Five-Year Plan for the Development of the Big Data Industry of Hubei Province	March 2022

2.3.3 *Planning Policy Tools*

The Chinese government adopted a variety of policy tools to promote and facilitate the development of relevant information industries. Taking the ICT related plans released from 2011 to 2020 as examples, total 90 plans or policy documents were collected and analyzed. Those documents were released by the General Office of the State Council, the MIIT, the National Development and Reform Commission (NDRC), the Ministry of Science and Technology (MST), and other departments. The commonly used policy tools were classified and counted according to the three aspects of supply, environment, and demand. The results are shown in Table 2.10.

The policy tools for the new generation of ICT industries were mainly supply-oriented and environment-oriented ones. During the 12th Five-Year Plan period, some subdivided policy tools were not activated, such as infrastructure construction within supply-oriented policy tools, and trade control within demand-oriented policy tools. Since the 13th Five-Year period, various types of policy tools have been enriched and expanded to a certain extent. In the supply-oriented policy tools, the talent development, public service, and technology investment were the main supporting tools, and their support increased significantly in 2016, while the support for infrastructure construction was relatively weak. In the environment-oriented policy tools, the government mainly promoted industrial development by formulating guiding policies like development goals and plans, followed by policies related to property rights protection and those related to financial support. In the demand-oriented policy tools, the most supportive tool was transformation of scientific and technological achievements, which grew rapidly after 2015, and had different trend compared with the supply-oriented policy tools. Now the Chinese government is actively exploring the application of different types of policy tools for industrial incentives and promotion methods, and promoting the development and progress of the industry considering the pains and shortcomings troubling various industries.

2.4 Development Goals and Key Areas

In February 2023, the CPC Central Committee and the State Council issued the "Overall Layout Plan for the Construction of Digital China". The plan pointed out that building "Digital China" is an important engine for promoting Chinese modernization in the digital era and a strong support for building new national competitive advantages. The construction of "Digital China" was laid out in accordance with the overall framework of "2522", namely, consolidating the "two foundations" of digital infrastructure and data resource system, and promoting the deep integration of digital technology and the Five-Sphere Integrated Plan for economic, political, cultural, social and ecological civilization construction, strengthening the "two major capabilities" of digital technology innovation system and digital security barrier, and

Table 2.10 Statistics on the policy tools adopted in China's ICT related plans (2011–2020)

Total types	Policy tools	Examples	Count in the selected set	Proportion of all tools
Supply-oriented	Talent development	Cultivate urgently needed specialized talents in R&D in key areas such as information technology	36	13.8%
	Infrastructure construction	Support the construction of electronic information industry engineering research centers, engineering laboratories, and enterprise technology centers	5	2.0%
	Public service	Support the construction of public R&D service platforms such as innovation knowledge centers, data centers, inspection and testing, and quality certification for the electronic information industry	32	12.3%
	Technology investment	Make full use of policy funds such as the National Integrated Circuit Industry Investment Fund to guide social capital investment and support the development of major industrialization projects	26	10.0%
Environment-oriented	Goal Planning	The construction of China's famous software city and national new industrialization industry demonstration base (software and information services) has reached a higher level, and the number of cities with industrial revenue exceeding 100 billion yuan has reached more than 20	47	18.0%

(continued)

Table 2.10 (continued)

Total types	Policy tools	Examples	Count in the selected set	Proportion of all tools
	Regulatory Control	Accelerate the establishment of industry standards and important product technical standards systems that are conducive to the development of the new generation of information technology industries, and optimize the approval and management procedures for market access	24	9.2%
	Open to the outside world	Encourage domestic integrated circuit enterprises to expand international cooperation, integrate international resources, and expand international markets	16	6.1%
	Financial support	Actively support qualified software and integrated circuit enterprises to raise funds through issuing stocks, bonds and other means to expand direct financing channels	11	4.2%
	Tax incentives	The VAT credits at the end of the period incurred by relevant integrated circuit enterprises due to the purchase of equipment will be refunded	15	5.7%
	Intellectual property protection	Actively develop and apply network copyright protection technology for genuine software to effectively protect software and integrated circuit intellectual property rights	17	6.5%

(continued)

Table 2.10 (continued)

Total types	Policy tools	Examples	Count in the selected set	Proportion of all tools
Demand-oriented	Government procurement	Implement a new generation of information technology industry innovative products and services promotion plan, and increase the government's first purchase and ordering efforts for software and other products	6	2.3%
	Transformation of scientific and technological achievements	Establish a benign interaction mechanism among enterprises, research institutes, and universities to provide full-process services to promote technology transfer and achievement transformation in the new generation of information technology industries	23	8.5%
	Trade controls	Support software exports and service outsourcing through export subsidies and other means	3	1.1%

optimizing the "two environments" (see Fig. 2.5) of domestic and international digital development.

The overall goal of "Digital China" is: by 2025, decisive progress will be made in the construction of "Digital China", the level of ICT development will be significantly upgraded, the digital infrastructure will be fully consolidated, the innovation ability of digital technology will be significantly enhanced, the value of data elements will be fully utilized, the digital economy will be developed in a high-quality way, and the efficiency of digital governance will be improved as a whole.

2.4.1 Digital Infrastructure

In March 2020, the Standing Committee of the Political Bureau of the CPC Central Committee clearly proposed to accelerate the construction of the new infrastructures such as 5G networks and data centers. The new infrastructures are those strategic and

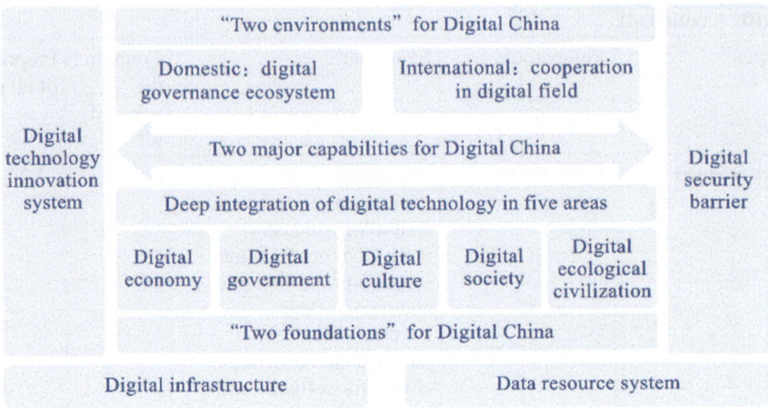

Fig. 2.5 Content framework of "Overall Layout Plan for the Construction of Digital China". *Image Source* China Netcom

network-based infrastructures with a multiplier effect, which are fit for the new generation of technological revolution and industrial transformation, and provide underlying support for the innovative, coordinated, green, open and shared development of the economy and society. There are three categories of new infrastructures. The first category is information infrastructure, which mainly refers to the infrastructure generated by the new-generation information technologies, such as the communications network infrastructures (e. g., 5G, IoT, Industrial Internet, and Satellite Internet), the computing power infrastructures (e. g., data centers and intelligent computing centers), and the new technology infrastructures (e. g., AI, cloud computing, and blockchain). The second category is integrated infrastructure, which mainly refers to the integrated infrastructure formed by deeply applying information and communication technologies like the Internet, big data, and AI to support the transformation and upgrading of traditional infrastructure. Examples of this category are the smart transportation infrastructure and smart energy infrastructure. The third category is innovative infrastructure, which mainly refers to public welfare infrastructure that supports scientific research, technology development, and product development. This category includes the major scientific and technological infrastructure, scientific and educational infrastructure, and industrial technology innovation infrastructure. With the technological revolution and industrial transformation, the connotation and extension of new infrastructures will not remain unchanged, but will continue to evolve and change. The main contents of China's new infrastructure are shown in Table 2.11.

In the "Overall Layout Plan for the Construction of Digital China" issued in February 2023, the "information infrastructure" in the "new infrastructures" was referred to "digital infrastructure". This document also presented the recent construction priorities. In terms of network infrastructure, accelerate the coordinated construction of 5G networks and gigabit optical networks, deepen the large-scale deployment and application of IPv6, promote the comprehensive development of mobile IoT, and vigorously promote the large-scale application of BDS. In terms of computing power

Table 2.11 Main contents of China's new infrastructure

Categories	Sub-categories	Scopes
Information infrastructure	Communications network infrastructure	5G, IoT, Industrial Internet, Satellite Internet, etc
	Computing power infrastructure	Data centers, intelligent computing centers, etc
	New technology infrastructure	AI, cloud computing, blockchain, etc
Integrated infrastructure	Smart transportation infrastructure	Intercity high-speed railway, urban rail transit
	Smart energy infrastructure	Ultra-high voltage transmission and transformation network, new energy vehicle charging piles
Innovative infrastructure	Public welfare infrastructure supporting the development of science and technology	Major scientific and technological infrastructure, scientific and educational infrastructure, industrial technology innovation infrastructure, etc

Table 2.12 The development goals of China's communications network infrastructure in the 14th Five-Year Plan period

Metrics	2020	2025
Internet user scale	989 million	1.2 billion
5G user penetration rate	15%	56%
Number of optical fiber users with 1000 Mbps or higher access bandwidth	6.4 million	60 million
Number of active IPv6 users	462 million	800 million

infrastructure, systematically optimize the layout of countrywide computing infrastructure, promote efficient complementarity and coordinated linkage of computing power in the east and west of China, and guide the reasonable deployment of data centers, supercomputing centers, intelligent computing centers, edge data centers, etc. In terms of application infrastructure, strengthen the digitalization and intelligent transformation of traditional infrastructure, and improve the intelligence level of social general infrastructure, such as electricity, transportation, and logistics. The development goals of China's communications network infrastructure in the 14th Five-Year Plan period are shown in Table 2.12.

2.4.2 *Digital Economy*

Since entering the 14th Five-Year Plan period, China has intensively released a number of important plans related to the digital economy, including the "Outline of the 14th Five-Year Plan for National Economic and Social Development and the

Table 2.13 The development goals of China's digital economy in the 14th Five-Year Plan period

Metrics	2020	2025
The added value of core industries in the digital economy as a percentage of GDP	7.8%	10%
In terms of digital industrialization, the scale of information consumption	5.8 trillion yuan	7.5 trillion yuan
In terms of digital industrialization, online retail value	11.76 trillion yuan	17 trillion yuan
In terms of industrial digitalization, the ratio of enterprise industrial equipment going cloud-ward	13.1%	30%
In terms of industrial digitalization, the proportion of enterprises with comprehensive digitalization of key business links	48.3%	60%

Long-Range Objectives Through the Year 2035 of the People's Republic of China", the "14th Five-Year Plan for Digital Economy Development", and the "14th Five-Year Plan for the Deep Integration of Informatization and Industrialization". The national 14th Five-Year Plan has established a systematic framework for promoting digital development. In terms of overall goals, it is proposed that by 2025, the added value of China's core industries in the digital economy will account for 10% of GDP, up from 7.8% in 2020. See Table 2.13 for China's digital economy development goals in the 14th Five-Year Plan period.

2.4.2.1 Digital Industrialization

Accelerate the promotion of digital industrialization. Cultivate and strengthen emerging digital industries such as AI, big data, blockchain, cloud computing, and cybersecurity, and improve the industrial level of communications equipment, core electronic components, and key software. Encourage enterprises to open up search, e-commerce, social and other data, and develop third-party big data service industries. Promote the sound development of sharing economy and platform economy. The key industries and development goals of China's digital industrialization in the 14th Five-Year Plan period are shown in Table 2.14.

2.4.2.2 Industrial Digitalization

Promote the transformation of industrial digitalization. Accelerate the integration and application of the new-generation ICT and the real economy, implement the "Using Data to Empower Wisdom on the Cloud" action, and create a smart supply chain system supported by big data, networked sharing, and intelligent collaboration. Build smart agriculture, accelerate the digitalization and intelligent upgrading of all links

Table 2.14 The key industries and development goals of China's digital industrialization in the 14th Five-Year Plan period

Fields	Development goals
Cloud computing	Accelerate the iteration and upgrading of cloud operating systems, promote technological innovations such as ultra-large-scale distributed storage, elastic computing, and data virtual isolation, and improve cloud security levels; focus on hybrid cloud to cultivate cloud service industries such as industry solutions, system integration, and operation and maintenance management
Big data	Promote technological innovations in big data collection, cleaning, storage, mining, analysis, and visualization algorithms, cultivate a full life cycle industry system for data collection, labeling, storage, transmission, management, and application, and improve the big data standard system
IoT	Promote technological innovations such as sensors, network slicing, and high-precision positioning, coordinate the development of cloud services and edge computing services, and foster the Internet of Vehicles (IoV), medical IoT, and home IoT industries
Industrial Internet	Build an independent and controllable identification resolution system, standard system, and security management system, strengthen the research and development and application of industrial software, cultivate and form an Industrial Internet platform with international influence, and promote the construction of the "Industrial Internet plus Intelligent Manufacturing" industrial ecosystem
Blockchain	Promote innovation in blockchain technologies such as intelligent contracts, consensus algorithms, encryption algorithms, and distributed systems, focus on developing blockchain service platforms and application solutions in the fields of financial technology, supply chain management, and government services with alliance chains as the key point, and improve regulatory mechanisms
AI	Build AI data sets for key industries, develop algorithm reasoning training scenarios, promote the design and manufacturing of intelligent products such as intelligent medical equipment, intelligent vehicles, and intelligent recognition systems, and promote the construction of general and industry-wide artificial intelligent open platforms

in the agricultural production, processing, sales, logistics and other industrial chains. Build an agricultural basic data resource system, and accelerate the informatization of agricultural science and technology services. Accelerate the digital transformation of the manufacturing industry, develop a multi-level systematic Industrial Internet platform system and innovative applications, strengthen the development of standard system for informatization and industrialization, and deeply implement the intelligent manufacturing projects. Deepen the digital transformation of the service industry and cultivate new growth points such as crowdsourcing design, smart logistics, and new retail. The typical industry development goals of China's industrial digitalization during the 14th Five-Year Plan period are shown in Table 2.15.

Table 2.15 The typical development goals for industrial digitalization in the 14th Five-Year Plan period

Industrial sectors	Development goals
Intelligent manufacturing	Promote equipment networking, digital connection of production links and coordinated response of supply chain, promote the integration of production data, flexible manufacturing, personalized products and intelligent management
Smart agriculture	Promote precision sowing, fertilization and pesticide application, and harvesting of field crops, and promote the intelligent application of facility horticulture, livestock and poultry aquaculture; build a smart water conservancy system, and improve water situation monitoring and intelligent scheduling capabilities based on river basins
Smart energy	Promote the intelligent upgrading of coal mines, oil and gas fields, power plants, etc., carry out extensive collection of energy consumption information and online analysis of energy efficiency, and realize the interaction of source, grid, load and storage, multi-energy synergy and complementarity, and intelligent regulation of energy demand
Intelligent transportation	Develop autonomous driving and vehicle–road collaborative travel services, promote intelligent highway management, traffic signal linkage, and public transportation priority control, and build smart railways, smart civil aviation, smart ports, digital waterways, and smart parking lots

2.4.3 Digital Society

The "14th Five-Year Plan for National Informatization" proposed to steadily advance the construction of a digital society. Cultivate new business models driven by digital technology and data resources. Encourage intelligent upgrading and business model innovation in the fields of travel, catering, accommodation, culture, tourism, sports, logistics, and housekeeping, promote brand consumption and quality consumption, and cultivate a high-quality digital life service market. The public service system should be more convenient and beneficial to the people, and the role of informatization in supporting basic livelihood security and basic social services should be effectively enhanced. The level of equalization of digital public services should be significantly improved, and the gap in service levels between urban and rural areas should be significantly narrowed. The development goals for typical fields of digital public services in China in the 14th Five-Year Plan period are shown in Table 2.16.

2.5 Summary

Under the leadership of the Communist Party of China, China has developed from a poor and backward agricultural country into a prosperous and progressive industrial country after more than 70 years of development. Earth-shaking changes have taken

Table 2.16 The development goals for typical fields of digital public services in the 14th Five-Year Plan period

Typical fields	Development goals
Smart education	Promote the inclusion of high-quality social online course resources into the public teaching system, promote the online radiation of high-quality educational resources to weak schools in rural and remote areas, and develop scenario-based, experiential learning and intelligent education management and evaluation
Smart healthcare	Improve electronic health records, medical records, electronic prescriptions and other databases, and accelerate data sharing among medical and health institutions; promote telemedicine, advance the application of medical image-assisted interpretation, clinical-assisted diagnosis, etc.; use big data to enhance the supervision of medical institutions and medical behaviors
Smart culture and tourism	Promote scenic spots, museums, etc. to develop online digital experience products, build scenic spot monitoring facilities and big data platforms, and develop new cultural and tourism services such as immersive experience, virtual exhibition halls, and high-definition live broadcasts
Smart community	Promote the interconnection of government service platforms, community sensing facilities and household terminals, develop community welfare services such as intelligent early warning, emergency rescue and smart elderly care, and establish an unmanned logistics distribution system
Smart home	Apply induction control, voice control, remote control and other technical means to develop smart home appliances, smart lighting, smart security monitoring, smart speakers, wearable devices, service robots, etc

place across China, and the Chinese people are enjoying prosperous and peaceful life.

China has not only built globally leading information infrastructure but also built a technologically advanced and comprehensively covered industrial chain, ranking at the forefront of the world in many application fields such as mobile Internet, e-commerce, and mobile payment. Currently, China's digital economy scale ranks second in the world in scale, and the proportion of added value from core digital economy industries in GDP has been gradually increasing. China's independent innovation capabilities in ICT have significantly improved, with scientific and technological strength and international competitiveness growing day by day.

As a socialist country, China demonstrates institutional advantages in industrial planning and development. The government has established an ICT industry and informatization planning system covering all industries, including multi-level layouts such as medium- and long-term plans, Five-Year plans, and short-term plans. It comprehensively uses policy tools such as supply-oriented, environment-oriented, and demand-oriented measures to accurately promote the development of the ICT industry. At present, China has established macro strategies such as "Cyber Power", "Smart Society", and "Digital China". "Cyber Power" focuses on infrastructure, key technologies, and application innovation; "Smart Society" promotes intelligent infrastructure and inclusive public services; "Digital China" emphasizes creating new advantages in the digital economy and fostering a new ecosystem for digital development.

The Chinese government attaches great importance to the transformative role of ICT in industries and society. Telecommunications infrastructure construction has been listed as a key task in several Five-Year plans, serving as the foundation to drive digital transformation and industrial upgrading across all industries. China is advancing steadily toward the goal of building a "Digital China", demonstrating strong momentum and a bright future in the global informatization.

Chapter 3
ICT Innovations

3.1 Scientific Research Innovation System

Deng Xiaoping, the chief architect of China's reform and opening-up, pointed out that science and technology are the primary productive forces. Information and communications technology (ICT) innovation is the core element in the development of the ICT industry. Due to the highly innovative and high-risk characteristics of the ICT industry, conducting organized scientific research has become a characteristic way for China to catch up with and surpass the world's advanced scientific and technological levels. This chapter first introduces China's existing scientific research innovation system, as well as the science and technology development plans. Then it focuses on China's major science and technology projects and industrial technology breakthrough achievements in the field of ICT in recent years.

When the People's Republic of China was founded in 1949, there were only around 30 institutions engaged in scientific and technological research nationwide, and there were almost no scientific research teams specializing in ICT. Since then, China gradually established ICT teaching, research and production bases from scratch. After about 70 years of development, China has established a relatively complete research and development system for ICT, including colleges and universities, scientific research institutions, industrial clusters and so on.

3.1.1 Strategic Scientific Research Force

In the 1950s, China initially established an independent ICT research and development team. In 1956, the Ministry of Education (MOE) gathered teachers and students from five universities to the Department of Physics of Peking University to establish the first semiconductor major in China. In 1956, the Institute of Computing Technology of the Chinese Academy of Sciences was established, and in 1960, the

W. Liu et al. (eds.), *Digital China*, Introduction to China's S&T Innovation,
https://doi.org/10.1007/978-981-95-6240-4_3

Institute of Semiconductors of the Chinese Academy of Sciences was established. After the reform and opening-up, the national strategic scientific and technological force system has been gradually improved, covering state key laboratories for basic research, and innovation bases for market applications such as the national engineering research centers and national technology innovation center.

3.1.1.1 State Key Laboratory

The state key laboratories are important scientific and technological innovation bases where China organizes basic research, applied basic research and cutting-edge technology research, gathers and cultivates leading innovative talents, conducts high-level innovative cooperation, and produces major original achievements. The state key laboratories that began to be established in 1984 had reached a total of 265 by 2021. The main supporting units are China's key universities and major scientific research institutions in related fields. The state key laboratories actively participated in international big science programs and big science projects, and promoted China's breakthroughs in basic theoretical research and major key technologies. By 2021, there were 26 state key laboratories in the field of ICT, involving semiconductors, electromagnetism, communications, computers, networks and other fields.

According to the "Outline of the 14th Five-Year Plan for National Economic and Social Development and the Long-Range Objectives Through the Year 2035 of the People's Republic of China", China will strengthen national strategic scientific and technological strength, promote the optimization and combination of innovation systems guided by national strategic needs, and accelerate the construction of strategic scientific and technological strength led by national laboratories. Focus on major innovation fields such as quantum information, photonics and micro-nano electronics, network communications, and artificial intelligence (AI) to establish a number of state laboratories, reorganize state key laboratories, and form a laboratory system with a reasonable structure and efficient operation. Optimize and upgrade innovation bases such as the national engineering research centers and the national technology innovation centers, carry out in-depth industry-university-institute cooperation, and promote the radiation and driving effect of basic theories and key technologies on the industry.

The Law on Scientific and Technological Progress of the People's Republic of China implemented in January 2022 proposed to "establish and improve a laboratory system led by state laboratories and supported by national key laboratories". National key laboratories have replaced state key laboratories to become important strategic scientific and technological forces. In July 2022, the Ministry of Science and Technology (MOST) organized a meeting to promote the optimization and reorganization of state key laboratories. After reorganization, recommendation and evaluation, the first batch of 20 benchmark national key laboratories were selected and approved for construction. Among them, there were 9 national key laboratories in the field of ICT, covering sub-fields such as AI, integrated circuits, and micro-nanoelectronics (see Table 3.1).

Table 3.1 Benchmark national key laboratories in the field of ICT (July 2022)

National key laboratories	Supporting institutions
National Key Laboratory of Micro-Nano Electronic Devices and Integrated Technology	Peking University
National Key Laboratory of Virtual Reality Technology and Systems	Beihang University
National Key Laboratory of Integrated Chips and Systems	Fudan University
National Key Laboratory of Brain-Computer Intelligence	Zhejiang University
National Key Laboratory of Autonomous Intelligent Unmanned Systems	Beijing Institute of Technology, Tongji University
National Key Laboratory of Cognitive Intelligence	iFLYTEK Co., Ltd., University of Science and Technology of China
National Key Laboratory of Processors	Institute of Computing Technology of the Chinese Academy of Sciences
National Key Laboratory of Integrated Circuit Materials	Shanghai Institute of Microsystem and Information Technology of the Chinese Academy of Sciences
National Key Laboratory of Multimodal Artificial Intelligence Systems	Institute of Automation of the Chinese Academy of Sciences

3.1.1.2 National Engineering Research Centers

The national engineering research centers (NERCs) are research and development entities organized by the National Development and Reform Commission (NDRC). These centers are established by institutions with strong research and development (R&D) capabilities and comprehensive strengths, including universities, scientific research institutes, and enterprises. They are an integral part of the national innovation system. In response to the needs of national major strategic missions and key construction projects, efforts are made to tackle key technologies and conduct experimental research, develop major equipment, and carry out engineering experimental verification of significant scientific and technological achievements, in order to break through the constraints of key technologies and core equipment.

As of the end of 2022, there were 191 national engineering research centers, and 50 of them were in the fields of ICT. The categories of supporting institutions are relatively wide, including well-known universities and scientific research institutions, major telecommunications service operators (such as China Telecom, China Mobile, China Unicom, etc.), major information technology companies (such as Baidu, iFLYTEK, etc.), major electronic manufacturing companies (such as BOE, Haier, etc.), and scientific and technological major forces which carry out informatization in various industrial sectors (such as China UnionPay, PetroChina, etc.). Table 3.2 lists 12 national engineering research centers in the fields of ICT.

Table 3.2 Typical national engineering research centers in the fields of ICT

National engineering research centers	Supporting institutions
NERC for Next Generation Internet Core Technologies	Tsinghua University
NERC for Mobile Private Networks	Beijing Jiaotong University
NERC for Mobile Internet Security Technology	Beijing University of Posts and Telecommunications
NERC for Next Generation Internet Access System	Huazhong University of Science and Technology
NERC for Optical Fiber Sensing Technology and Network	Wuhan University of Technology
NERC for New-Generation Mobile Communications Test and Verification	China Academy of Information and Communications Technology
NERC for Cybersecurity Emergency	National Computer Network and Information Security Management Center
NERC for Cloud Network Infrastructure Security	China Telecom Global Limited
NERC for Next-Generation Internet Broadband Service Application	China Unicom Network Technology Research Institute
NERC for New-Generation Mobile Communications Technology	China Mobile Communications Group Co., Ltd
NERC for Mobile Communications and Internet of Vehicles	Datang Telecom Technology Co., Ltd
NERC for Digital Home Network	Haier Group Co., Ltd

According to the "Outline of the 14th Five-Year Plan for National Economic and Social Development and the Long-Range Objectives Through the Year 2035 of the People's Republic of China", China will concentrate on integrating and upgrading a number of key common technology platforms, and support leading enterprises in the industry to cooperate with universities, scientific research institutes and upstream and downstream enterprises in the industry to jointly build national industrial innovation centers, which will undertake the major national scientific and technological projects. China will support qualified enterprises to transform into scientific research institutes, and provide public welfare and common technical services. Some common technology platforms will be set up to solve key common technical problems across industries and fields.

3.1.2 High-Tech Industrial Development Zones

High-tech industrial development zones (HIDZs) are specific areas established with the approval of governments at all levels for the purpose of developing high-tech.

They mainly rely on the advantages of being knowledge-intensive and technology-intensive, as well as their regional economic strength, geographical location and openness to the outside world, and focus on the development of high-tech industries supported and encouraged by national policies. Enterprises settled in the HIDZs can enjoy preferential policies such as income tax preferential tax rate, export tariff exemption, and plant construction tax exemption. By the end of 2020, the total number of national high-tech industrial development zones (NHIDZs) had reached 169, including 70 in the east, 44 in the central region, 39 in the west, and 16 in the northeast; 21 national independent innovation demonstration zones (NIIDZs) had been established and become important carriers for implementing the innovation-driven development strategy. These zones had become the "leaders" of local innovation and development. From 2015 to 2020, the GDP of HIDZs increased from 8.1 trillion yuan to 12.2 trillion yuan, an increase of more than 50% in five years. The number of high-tech enterprises increased from 79,000 to 225,000.

The HIDZs have become an important scientific and technological force for breakthroughs in key technologies and enhancement of original innovation capabilities. In 2021, the NHIDZs gathered 84% of the state key laboratories and 78% of the national technology innovation centers. Enterprise R&D spending was 3.3 times that of 2012, accounting for 50% of the country's total; "Patent Cooperation Treaty" (PCT) international patent applications accounted for about 50% of the country's total. By the end of 2022, there had been 128 innovative industrial clusters in the NHIDZs. The scale of information technology industry in Beijing Zhongguancun NHIDZ accounted for 17% of the country's total; the scale of optoelectronic information industry in Wuhan East Lake NHIDZ accounted for 50% of the country's total; the scale of integrated circuit industry in Shanghai Zhangjiang NHIDZ accounted for 35% of the country's total. The HIDZs had cultivated and gathered 36.2% of the country's high-tech enterprises, 67.1% of the companies listed on the Science and Technology Innovation Stock Board, and 35.9% of the technology-based small and medium-sized enterprises. A number of world-class enterprises had also emerged.

3.1.3 Innovative Industrial Clusters

Since the founding of the People's Republic of China, China has preliminarily established an independent ICT industry chain, covering upstream and downstream links from raw materials to manufacturing of devices, instruments, and system equipment, and formed a regional industrial cluster with national information security industry bases and national information industry parks as the main body, mainly concentrated in the three major regions of the Bohai Rim Region (Beijing/Tianjin), the Yangtze River Delta Region (Shanghai/Hangzhou), and the Pearl River Delta (Guangzhou/Shenzhen).

In recent years, China has regarded ICT industry as one of the strategic emerging industries, and promoted it in combination with industrial cluster policies. In 2010, the "Decision of the State Council on Accelerating the Fostering and Development of

Strategic Emerging Industries" was issued, and the new-generation information technology was listed as one of the strategic emerging industries. In 2019, the "Notice of the National Development and Reform Commission on Accelerating the Construction of Strategic Emerging Industry Clusters" was issued, announcing the first batch of national strategic emerging industry cluster construction lists in 12 key fields, involving 66 clusters in 22 provinces, autonomous regions, and municipalities, of which 23 industrial clusters were established in the field of new-generation information technology, as shown in Table 3.3. In order to promote the balanced development of the information technology industry in various regions, Shanghai located in the eastern China, Wuhan and Hefei located in the central China, each has three industrial clusters on the list, becoming the two cities with the most strategic emerging industry clusters in the field of new-generation information technology.

The industrial cluster has the characteristics of network organization and flexible manufacturing, which is conducive to the realization of continuous technological innovation and the promotion of regional economic development. The "Outline of the 14th Five-Year Plan for National Economic and Social Development and the Long-Range Objectives Through the Year 2035 of the People's Republic of China" and the "14th Five-Year Digital Economy Development Plan" clearly stated that localities with conditions are encouraged to establish mixed-ownership industrial technology research institutes based on industrial clusters to serve the research and development of key common technologies in the region, actively explore the joint operation mode of platform enterprises and industrial parks, and explore the establishment of a new mechanism for cross-regional and cross-platform collaboration of various industrial clusters.

3.1.4 Technology Innovation Capability

According to the Global Innovation Index 2021 released by the World Intellectual Property Organization (WIPO), China ranked 12th in the world, went up two places from 2020, and was the only middle-income economy in the top 30 of the world, ranking higher than the developed economies such as Japan, Israel, and Canada. In the world's top 100 science and technology clusters of 2021, China had 19 clusters on the list, only 5 fewer than the United States (US), ranking second. Among them, the cluster of Shenzhen-Hong Kong-Guangzhou ranked second, and that of Beijing rose to third. In 2021, China's total number of PCT international patent applications was 69,540, ranking first in the world for three consecutive years; the number of PCT international patent applications in the field of ICT exceeded 30,000, an increase of 60% from 2017, accounting for 37.80% of the world's total (see Fig. 3.1).

In 2021, China ranked first in the world in the number of PCT international patent applications in the fields of computer technology and digital communications, accounting for more than one-third of the world; the two fields with the fastest growth in the effective number of domestic invention patents in China were information technology management methods and computer technology, with a year-on-year

Table 3.3 Industrial clusters established by China in the field of new-generation information technology

Subfields	Names
Integrated circuit	Shanghai Pudong New Area Integrated Circuit Industry Cluster
	Xi'an Integrated Circuit Industry Cluster
	Beijing Economic-Technological Development Area Integrated Circuit Industry Cluster
	Wuhan Integrated Circuit Industry Cluster
	Hefei Integrated Circuit Industry Cluster
New display device	Hefei New Display Device Industrial Cluster
	Shenzhen New Display Device Industrial Cluster
	Wuhan New Display Device Industrial Cluster
Next-generation information network	Wuhan Next-Generation Information Network Industry Cluster
	Yingtan Next-Generation Information Network Industry Cluster
	Zhengzhou Next-Generation Information Network Industry Cluster
Information technology services	Hangzhou Information Technology Service Industry Cluster
	Jinan Information Technology Service Industry Cluster
	Guiyang Information Technology Service Industry Cluster
	Dalian Information Technology Service Industry Cluster
	Shanghai's Yangpu District Information Technology Service Industry Cluster
	Chengmai Information Technology Service Industry Cluster
	Zhengzhou Information Technology Service Industry Cluster
Network information security products and services	Tianjin Binhai High-Tech Industrial Development Area Network Information Security Product and Service Industry Cluster
Artificial intelligence	Beijing Haidian District Artificial Intelligence Industry Cluster
	Hefei Artificial Intelligence Industry Cluster
	Shanghai Xuhui District Artificial Intelligence Industry Cluster
	Shenzhen Artificial Intelligence Industry Cluster

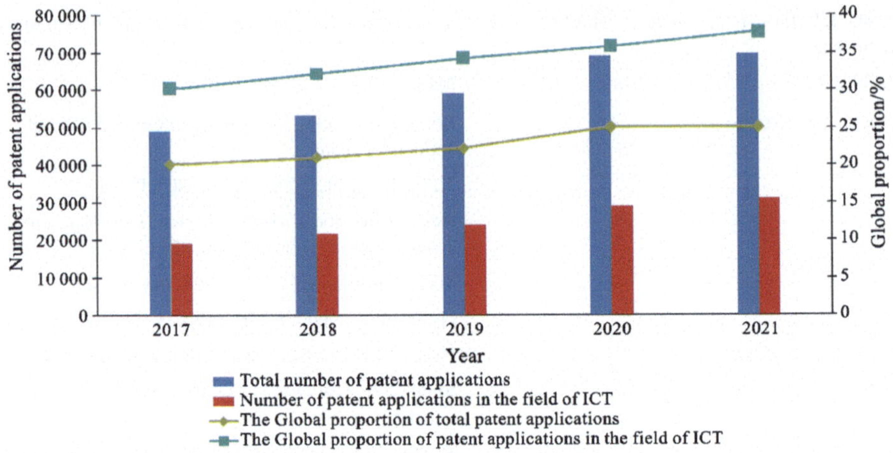

Fig. 3.1 Growth of China's PCT international patent applications (2017–2021). *Data Source* World Intellectual Property Organization

increase of 100.3% and 32.7% respectively; the innovative vitality of enterprises had been further stimulated, 13 Chinese enterprises had entered the top 50 list of PCT international patent applicants in the world, and Huawei had topped the list for five consecutive years. The number of high-level papers in the field of digital technology had further increased. As of September 2021, the number of highly cited papers in China was 42,900, accounting for 24.8% of the global share, among which the number of international citations of papers in the field of computer science had risen to the first place in the world.

China has made great strides in fields of 5G, AI, and other technologies. As of October 2020, China had declared a total of 13,282 5G standard essential patents, accounting for 37% of the global total. Huawei and ZTE ranked among the top 3 in the world in the 5G standard-essential patent declarations ranking published by the European Telecommunications Standards Institute (ETSI). China had become the second largest AI innovation center in the world after the US Among the top 100 highly cited papers in the direction of AI in the world from 2015 to 2020, China produced 21 papers, ranking second. From 2011 to 2020, China's patent applications in the field of AI accounted for 74.7% of the global total, 8.2 times that of the US.

In addition, China has made new progress in many core technologies in the fields of ICT. The development of key basic software has been accelerating, the ecological compatibility of desktop operating systems has continued to improve, the distributed operating system for all scenarios has entered the stage of industrialization, and some technical indicators of cloud databases are leading the world. The chip design capabilities in some fields have reached the internation top tier, the development of three dimensions non-volatile flash memory (3D NAND flash memory), dynamic random access memory (DRAM), and other memory technologies has been accelerating, and tens of millions of gate-level field programmable gate array (FPGA) products have been successfully mass-produced. The new display industry has made new breakthroughs. In 2020, the TFT-LCD production capacity in Chinese mainland ranked first in the world, and domestic flexible active matrix organic light emitting diode (AMOLED) entered the supply chain of international first-tier brands. The key technical capabilities of optical communications continue to improve, and some high-end optoelectronic chips such as laser chips, detector chips, and supporting electrical chips above 25G have been mass-produced.

3.2 Science and Technology Development Planning

3.2.1 National Science and Technology Plans

Since China introduced its first national science and technology plan in 1982, there have been more than 30 science and technology plans that support the national economy and social development and the advancement of science and technology from different angles. These plans complemented each other and formed an organic whole. China's current national science and technology plan system can be divided into five categories, namely the National Natural Science Foundation of China (NSFC), the National Science and Technology Major Project, the National Key Research and Development Program, the Technological Innovation Guidance Project, the Base and the Talent Project. Among them, the first three categories are the main traction plans for scientific and technological innovation.

1. National Natural Science Foundation of China. This is a nationwide natural science foundation established to promote the reform of China's science and technology system and change the way of scientific research funding, aiming to support national basic research. During its development, the NSFC gradually formed a funding structure consisting of three series of projects, namely research projects, talent projects and environmental conditions projects, promoting the development of China's natural science research and promoting the construction of basic disciplines, and discovering and cultivating outstanding scientific and technological talents.
2. National Science and Technology Major Project. To achieve national goals, these major projects were designed to produce major strategic products and key common technologies through core technology breakthroughs and resource integration, in certain time limit. By giving full play to the advantages of the socialist system in concentrating forces on major tasks and the role of market mechanisms, national science and technology major projects strive to make breakthroughs, achieve leapfrog development of productivity driven by partial leaps in technological developments and fill in the gaps in national strategies.
3. National Key Research and Development Program. Facing the world's scientific and technological frontiers, the main battlefield of the economy, and major national needs, these programs focus on funding major social public welfare

Table 3.4 Representative ICT projects supported by national science and technology plans of China (2020–2022)

National science and technology plans	Representative projects
National Natural Science Foundation of China	Major Research Plan "Basic Theory and Key Technology Research of Inclusive Robots", Major Research Plan "Basic Theory and Key Technology of Future Industrial Internet", Major Research Plan "Construction and Manipulation of the Second-Generation Quantum System"
National Science and Technology Major Project	Core Electronic Devices, High-end General Chips and Basic Software Products, Very Large-Scale Integrated Circuit Manufacturing Technology and Complete Sets of Processes, New-Generation Wireless Mobile Communications
National Key Research and Development Program	New-Generation Artificial Intelligence, Advanced Computing and Emerging Software, Social Governance and Smart Society Technology Supports, Intelligent Sensors, Intelligent Robots, Cyberspace Security Governance, Information Photon Technology, Micro-Nano Electronic Technology, Multi-Modal Network and Communications, Blockchain, New Display and Strategic Electronic Materials, Quantum Control and Quantum Information

researches that require long-term evolution in the fields related to the national economy and People's livelihood, such as agriculture, energy resources, ecological environment, and health, and strategic, fundamental and forward-looking major scientific issues, major common key technologies and products, and significant international scientific cooperation related to the core competitiveness of industries, overall independent innovation capabilities and national security.

The ICT related fields are the key areas supported by the national science and technology plans. The ICT related projects currently supported by the National Natural Science Foundation of China, the National Science and Technology Major Project, and the National Key Research and Development Program are shown in Table 3.4.

3.2.2 *National Digitization Projects*

According to the "14th Five-Year Plan for National Informatization", China divided the construction goals of Digital China into multiple goals, including digital infrastructure system, data element resource system, digital productivity innovation and development system, digital industry system, industry digital transformation and development system, digital social governance system, etc. The plan set up 17 national key informatization projects, as shown in Table 3.5.

Table 3.5 National key informatization projects in the 14th Five-Year Plan period

Fields	Project names
Digital infrastructure	5G Innovative Application Engineering
	"Intelligent Network Connection" Facility Construction and Application Promotion Project
	National Integrated Big Data Center System Construction Project
	Three-dimensional Network Construction and Application Demonstration Project of Space, Sky, Land and Sea
Digital technology innovation system	Core Technology Breakthrough Project in the Field of Information
	Information Technology Intellectual Property and Standardization Innovation Project
Digital economy	Data Element Market Cultivation Project
	Big Data Application Improvement Project
	Information Technology Industry Ecological Cultivation Project
	Manufacturing Digital Transformation Engineering
	Information Consumption Expansion and Quality Improvement Project
Digital society	Smart Public Security Development and Improvement Project
	Artificial Intelligence Social Governance Experimental Project
	Emergency Management Modernization Capacity Improvement Project
Digital government	National Integrated Government Service Improvement Project
Digital livelihood	Digital Public Service Optimization and Upgrading Project
Internaotional cooperation in the digital field	"Digital Silk Road" Joint Construction and Shared Use Project

3.2.3 R&D Investment

Since surpassing Japan in 2009 and the European Union (EU) in 2013, China has been the world's second-largest R&D investment spender after the US. In 2019, China's R&D investment exceeded the 2 trillion yuan mark for the first time; in 2020, R&D investment reached 2.4 trillion yuan, and the R&D investment intensity (ratio to GDP) was 2.4%. Calculated in 2019 purchasing power parity dollars, China's R&D investment accounted for about 25% of the total R&D investment of 37 countries counted by the Organization for Economic Cooperation and Development (OECD), second only to the US (about 30%). According to the "Outline of the 14th Five-Year Plan for National Economic and Social Development and the Long-Range Objectives Through the Year 2035 of the People's Republic of China", during the 14th Five-Year Plan period, China will increase fiscal investment in basic research, optimize the investment structure, implement tax incentives for enterprises to invest in basic research, encourage the society to invest in multiple channels such as donations and the establishment of funds, and form a sustained and stable investment mechanism. The proportion of basic research investment in R&D investment will increase to more than 8%.

In addition to the national investment in basic research and public information facilities, industry enterprises also invested a huge amount of R&D funds. The ICT industry is a typical technology-intensive industry with rapid technology iteration and rapid market changes. Enterprises often need to invest higher R&D costs to quickly iterate related technologies and seize technological opportunities and commanding heights. From 2017 to 2021, the R&D investment of listed Internet enterprises in China increased by 227%. Among the global top 2500 industry enterprises in R&D investment released by the European Commission in 2021, 597 Chinese enterprises were selected, including 210 enterprises in the fields of information technology software services and hardware equipment. For example, Huawei's R&D investment in 2012 was 30.1 billion yuan, with an intensity of 13.7%, and its R&D investment in 2021 increased to 142.7 billion yuan, with an intensity of 22.4%; ZTE's R&D investment in 2012 was 8.829 billion yuan, with an intensity of 10.5%, and its R&D investment in 2021 was 18.8 billion yuan, with an intensity of 16.4%. The overall R&D investments of these enterprises are much higher than those of traditional enterprises and general manufacturing enterprises, laying the foundation for the rapid development of ICT.

3.3 Major Science and Technology Project

When a nation faces major opportunities and challenges in its development, it is essential to prioritize national interests, follow the guidance of national will, and accurately grasp strategic objectives. This requires pooling the nation's top scientific, technological, and management talents to undertake foundational major science and technology projects. Such a whole-nation system has been widely adopted globally for technological innovation, exemplified by the US Manhattan Project and Apollo Program, Japan's Fifth Generation Computer System (FGCS) project and Very Large Scale Integration (VLSI) project. This section highlights China's advancements in basic research and critical infrastructure through three case studies: supercomputers, communications and quantum computing, and the BeiDou Navigation Satellite System (BDS).

3.3.1 Supercomputer

A supercomputer refers to a computer that could perform a large number of high-speed calculations that ordinary personal computers cannot handle. High-performance computing has become the evolution direction of data-intensive research disciplines such as high-energy nuclear physics, space science, material science, life science, and AI. It is the core infrastructure that supports industrial scale growth and application model innovation. It is the strategic high ground that major developed countries in the world are vying for. The US, the EU and Japan have successively formulated national-level plans such as open science network, high-performance infrastructure, and advanced computing partners.

China's supercomputer research started in the 1970s. In October 1972, the development of supercomputers was listed into the national key projects. In December 1983, China developed the first exascale supercomputer "Galaxy". It became the computer system with the fastest computing speed, the largest storage capacity, and the most powerful functions in China at that time, breaking the technological monopoly of developed countries in this field. China became the country that could independently design and manufacture supercomputers after the US, Japan, France, Britain, and Germany. After the "Galaxy-I", China has developed a series of "Galaxy" supercomputers of different sizes.

In 2006, the MOST of China listed high-performance computers as part of the national science and technology plan. "Tianhe-1", as a major project of China's "High-tech Research and Development Plan" (863 Plan), was developed on the basis of the "Galaxy" series of supercomputers of the National University of Defense Technology (NUDT). On October 29, 2009, the NUDT successfully developed the "Tianhe-1" supercomputer with a peak performance of 1206 trillion calculations per second, making China the second country in the world capable of developing petaflop supercomputers after the US. On June 20, 2016, the International Supercomputing Conference was held in Frankfurt, Germany. During the conference, the 47th list of the International TOP 500 Organizations was officially released. China's "Sunway Taihu-Light" replaced "Tianhe-2" to top the list (see Fig. 3.2). In the top 10 of the 2022 International TOP 500 list, China had two supercomputers, namely "Sunway Taihu-Light" ranked sixth and "Tianhe-2" ranked ninth. The former had a peak computing power of 125.44 quadrillion times per second, and the latter had a peak computing power of 100.68 quadrillion times per second. In the 2022 International TOP 500 list, China continued to lead the world with 186 supercomputers, and the US ranked second in the world with 123 supercomputers.

"Sunway Taihu-Light" used the chips with China's independent intellectual property rights, which are faster and more energy-efficient. A "chip" of about five square centimeters integrates 260 computing cores, billions of transistors, and has a computing power of more than 3 trillion operations per second. The computing power of the "Sunway Taihu-Light" supercomputer for 1 min is about 32 years of non-stop calculations by more than 7 billion people around the world at the same time. At present, China's supercomputer applications involve weather and climate,

Fig. 3.2 "Sunway Taihu-Light" supercomputer (2016). *Image Source Shenzhen Special Zone Daily*

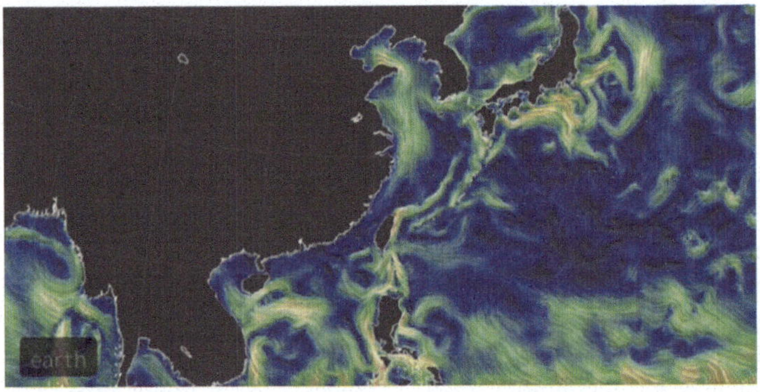

Fig. 3.3 Using supercomputers to forecast marine environmental elements. *Image Source* National Supercomputer Center in Guangzhou

earth science, marine environment, material design, biopharmaceuticals, aerospace, graph computing, quantum computing, life science, astrophysics and many other areas (see Fig. 3.3). For example, by using supercomputers to calculate the high-resolution China Sea regional atmosphere–ocean coupling model, forecasts of three-dimensional ocean currents, waves, sea temperature, salinity, sea surface height and other marine environmental elements for the next 10 days can be completed within 25 min.

3.3.2 Quantum Communications and Quantum Computer

Quantum computer was first proposed by American physicist Feynman in 1982. It is a physical device that uses the principles of quantum mechanics to perform high-speed mathematical and logical operations, and store and process quantum information. The natural parallel characteristics of quantum computing will play a huge role in processing certain large-scale parallel operations, and play a key role in the fields of massive information processing and research on major scientific issues. It will have a significant impact on one country's economy, science and technology, and military and information security, and has become one of the scientific and technological focuses of common concern and exploration by the global human race. In 2018, Google Inc. released a 72-bit superconducting quantum chip. In 2019, IBM of the US released the IBM Q System One quantum computer. In 2020, the US announced the construction of five quantum science joint research centers, including the Next Generation Quantum Science and Engineering (Q-NEXT), to continue to promote the development of quantum technology.

In 2005, China listed "Quantum Control Research" as one of the national basic research program of China in the "Outline of National Medium-and Long-Term Science and Technology Development Plan (2006–2020)". In 2017, China listed "Quantum Communications and Quantum Computer" in the "Science and Technology Innovation 2030-Major Project" and provided strong support. In August 2016, China successfully launched the world's first quantum science experiment satellite "Mozi" into space with the Long March 2D carrier rocket at the Jiuquan Satellite Launch Center. The research team of the University of Science and Technology of China (USTC) established quantum entanglement and secure communications between two ground stations more than 1120 km apart through "Mozi"(see Fig. 3.4), and published a research paper titled "Entanglement-based secure quantum cryptography over 1120 km" online in the journal Nature in June 2020. According to China's quantum communications development plan, after the launch of the quantum satellite, China will build a "quantum communications trunk line between Beijing and Shanghai", and a wide-area quantum communications system will initially be formed in the country. By around 2030, China will take the lead in building a global quantum communications network.

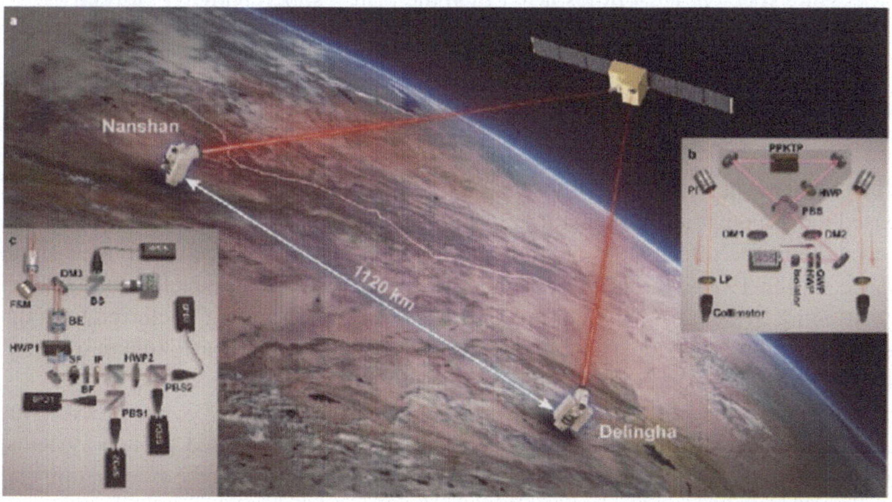

Fig. 3.4 China's quantum science experimental satellite "Mozi" achieves long-distance ground quantum communications. *Image Source* University of Science and Technology of China

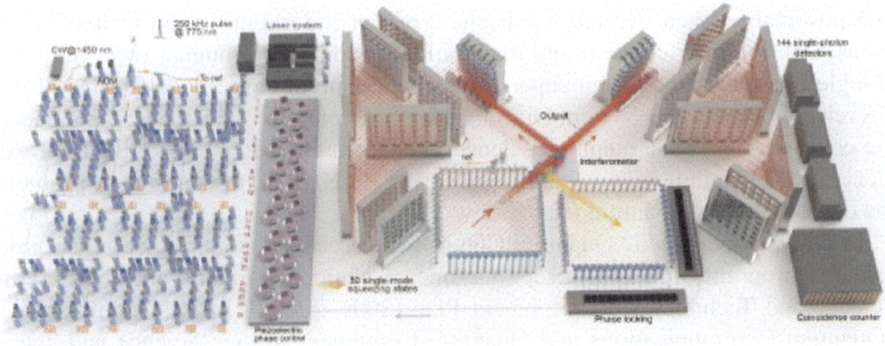

Fig. 3.5 China's quantum computing prototype "Jiuzhang-2" was successfully developed. *Image Source* University of Science and Technology of China

In 2017, Prof. Pan Jianwei's team at USTC built the world's first optical quantum computing prototype that surpasses early classical computers, and gradually demonstrated the superiority of quantum computing through experiments. In 2020, the team successfully built a Gaussian boson sampling quantum computing prototype named "Jiuzhang" with 76 photons and 100 modes. The output quantum state space scale of "Jiuzhang" reached 1030, and its speed of processing Gaussian boson sampling was 100 trillion times faster than the supercomputer "Fugaku". In 2021, the number of photons detected by "Jiuzhang-2"(see Fig. 3.5) increased to 113, and the output state space dimension reached 1043. It completed the rapid solution of the Gaussian boson sampling task used to demonstrate the "superiority of quantum computing", and the processing speed is 1024 times faster than the fastest supercomputer currently.

3.3.3 BeiDou Navigation Satellite System

The global navigation satellite system (GNSS) can provide users with all-weather air-based radio navigation and positioning services with three-dimensional coordinates, speed, and time information at any location on the earth's surface or in near-Earth space. In the 1970s, the US Department of Defense began to develop the "Global Positioning System for Navigation Satellite Timing and Ranging", referred to as the global positioning system (GPS). The GPS plays an important basic role in military, disaster prevention and mitigation, and surveying and mapping.

In 1994, China initiated the construction of the BDS. In 2003, the BDS-1 was built, consisting of three geostationary orbit satellites and related ground systems, which solved the problem of the existence of China's satellite navigation system and realized the active positioning of China and surrounding areas. China also became the third country in the world to have an independent satellite navigation system after the US and Russia.

In 2004, China launched the construction of the BDS-2 system with the goal of establishing a regional satellite navigation system. In December 2011, the BDS started to trial operation services, and provided services to China and parts of the Asia–Pacific region. The BDS adopted a hybrid constellation solution consisting of geostationary orbit, inclined geosynchronous orbit, and medium earth orbit satellites, realizing the integrated active and passive navigation multifunctional services.

In 2012, China launched the construction of the BDS-3 system. The research team conquered the constellation inter-satellite link technology, adopted an integrated design of inter-satellite and satellite-to-ground transmission functions, and realized the link interconnection between satellites and satellites, and between satellites and ground stations. On June 23, 2020, the 55th BeiDou navigation satellite was successfully launched, and the constellation deployment of the BDS-3 GNSS was fully completed. On July 31, 2020, Chinese President Xi Jinping announced the official commissioning of the BDS-3 GNSS system. The BDS is shown in Fig. 3.6.

The "White Paper on the Development of China's Satellite Navigation and Location Services Industry (2022)" showed that in 2021, the core output value of BeiDou-related industries, including chips, devices, and algorithms directly related to the research and development and application of satellite navigation technology, reached 145.4 billion yuan, and the associated output value driven by satellite navigation applications and services reached 323.6 billion yuan. The total number of enterprises and institutions in the field of satellite navigation and location services reached 14,000, and the number of employees exceeded 500,000. As of January 2023, the global cumulative number of intelligent device accessing to BeiDou space–time intelligent services exceeded 1.5 billion; BeiDou high-precision space–time services were called more than 170 billion times per month, and the service covered more

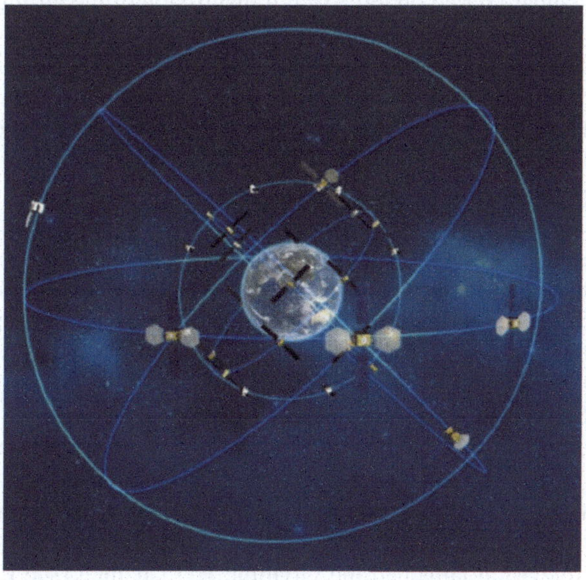

Fig. 3.6 Schematic diagram of BDS. *Image Sources Guangming Daily*

than 230 countries and regions around the world; BeiDou had become the standard configuration of Chinese demestic smartphones. For example, the mobile APP of AutoNavi Maps called the positioning services of BeiDou satellites for more than 300 billion times per day.

3.4 Industrial Technology Innovation

The development of many key technologies in the ICT industry requires not only huge capital investment, but also clear market returns. China integrates the power of universities, scientific research institutions and enterprises in various ways to jointly break through the technical bottleneck of the information industry and promote industrial progress and development. This section selects some typical industries such as ultra-HD video, mobile communications, and integrated circuits to introduce China's progress in technological breakthroughs in the ICT industry.

3.4.1 Ultra-HD Video Industry

High-definition television (abbreviated as HDTV) was firstly developed and tested in 1968. The Japan Broadcasting Corporation (NHK) completed the research and development of the HDTV system and the MUSE satellite transmission system around 1980, and released the first HDTV video signal standard—1125/60 standard. In 1996, the Advanced Television Systems Committee (ATSC) of the US released the HDTV standard based on terrestrial transmission of the all-digital system.

China's attention to HDTV began during the 8th Five-Year Plan period. In 1993, the National Science and Technology Commission organized and established the "HDTV Development Strategy Expert Group". In 1998, China's first terrestrial digital HDTV prototype was successfully developed. In October 2003, the NDRC established a terrestrial digital television transmission standard research and development project. Tsinghua University, Shanghai Jiao Tong University, Academy of Broadcasting Science and other eight universities and research institutions established a special working group on the national standard for terrestrial transmission of digital television, aiming to develop a new solution based on the three standard solutions of Tsinghua University's DMB-T, Shanghai Jiao Tong University's ADTB-T, and the Academy of Broadcasting Science's TiMi. In August 2006, "Framing Structure, Channel Coding and Modulation for Digital Television Terrestrial Broadcasting System" became a mandatory national standard, with the standard number "GB 20600-2006", which was officially implemented on August 1, 2007.

At 9:00 a.m. on January 1, 2008, the China Central Television (CCTV) launched free service of terrestrial digital HDTV broadcasting, marking the substantive launch of China's national standard for terrestrial digital television. On May 1, 2008, Beijing Television launched the Olympic HD Channel (see Fig. 3.7). The core equipment used in this batch of HD channels was all domestically produced, and basically all adopted the terrestrial digital TV single-carrier mode and related equipment. The development of HDTV in China took advantage of the opportunity of the Olympic Games and achieved good results. By the end of 2008, China had 45.28 million digital TV users, a year-on-year increase of nearly 70% over 2007, of which more than 500,000 were HDTV users, and Beijing users accounted for more than 60%. By March 2009, 229 cities in China had completed the overall digital conversion of cable TV.

HDTV technical standards have promoted the upgrading of the TV industry. In June 2012, the starting meeting of the 12th Five-Year Plan 863 Program "Research and Verification of New-Generation Digital TV Key Technologies" was held in Beijing to promote "synergy between content and network", "interaction between broadcasting network and Internet", "new wireless overlay network design" and other technological breakthroughs. In February 2019, the Ministry of Industry and Information Technology (MIIT), the National Radio and Television Administration, and CCTV jointly issued the "Ultra-HD Video Industry Development Action Plan (2019–2022)", which clarified the overall requirements, development goals, key tasks and safeguards for the development of the Ultra-HD video industry. In 2019, China's Ultra-HD video display technology gradually matured, mainly including memory chips, high-performance display driver chips, TV main control chips, Ultra-HD set-top box chips, 4 K/8 K image processing chips, etc. The sales of 4 K TVs in China were gradually increasing, occupying a dominant position in the Ultra-HD video industry.

The development of the 4 K TV industry has promoted the gradual maturity of the 4 K Ultra-HD video industry ecosystem. The market share and international influence of well-known Chinese companies (such as TCL, Hisense, Skyworth, etc.) continue to increase. In 2019, the new set-top boxes of China's three major telecommunications operators fully supported 4 K 60 frames/HDR/BT.2020/10bit video decoding capabilities. In 2020, the global shipment of 4 K set-top boxes exceeded 120 million units, and China became the world's largest market for 4 K set-top boxes.

Fig. 3.7 The HD broadcasting studio of Beijing Television during the 2008 Beijing Olympic Games. *Image Source* CCTV.com

3.4.2 Mobile Communications Industry

The mobile communications technology involves the information infrastructure of various countries, and the competition for the new generation of 5G mobile communications technology is fierce. Internationally, South Korea launched the formulation of South Korea's national 5G standards in 2017 and launched 5G commercial services in 2019; the EU issued the "5G Action Plan" in 2016, and announced in 2017 that it would allocate 700 million euros in the "Horizon 2020" scientific research plan to support the development of 5G; the US released the 24 GHz high-frequency band of 5G in July 2016, and planned to invest 400 million USD to support 5G trials and research and development.

China's mobile communications industry experienced the development process of "following in 2G", "breakthrough in 3G", and "synchronization in 4G", and are striving to achieve the new goal of "leading in 5G". In the 2G era, China introduced foreign 2G technical standards such as the global system for mobile communications (GSM) and code division multiple access (CDMA), and gradually realized independent research and development of equipment and products. In the 3G era, China's time division-synchronous code division multiple access (TD-SCDMA) standard was approved by the International Telecommunication Union (ITU) in May 2000 as one of the 3G international standards, and was successfully commercialized in 2009. In the 4G era, the time division-long term evolution-advanced (TD-LTE-Advanced) standard led by China became the 4G international standard in January 2012.

With the technical accumulation and industrial achievements of 3G and 4G, China fully launched the work of 5G research and development and standardization. In February 2013, the MIIT, the NDRC, and the MOST jointly promoted the establishment of the IMT-2020 (5G) Promotion Group to coordinate China's 5G technology research, standard formulation, R&D trials, and industrial promotion. In October 2015, the ITU's 2015 World Radiocommunication Conference was held in Geneva, Switzerland. At this conference, eight of the nine technical indicators of the "Flower of 5G" proposed by China were adopted by the ITU (see Fig. 3.8). Since then, key technologies such as flexible system design, polarization codes, large-scale antennas and new network architecture proposed by Chinese enterprises have become the focus of international standards. Chinese technical experts have began to hold multiple important positions in international standard organizations such as the ITU and the Third Generation Partnership Project (3GPP), and preside over related work on key projects.

In the field of 5G technology research and development, China's scientific research teams and enterprises have made significant breakthroughs. In terms of 5G baseband chips, Huawei's multi-mode commercial 5G baseband chip Balong 5000 supports 2G/3G/4G/5G networks, Sub-6 GHz, millimeter wave, 5G non-standalone networking (NSA) and 5G standalone networking (SA). In the field of 5G air interface technology, Huawei and ZTE are leaders in Massive MIMO technology, and Huawei released the world's first 5G base station core chip. In the field of channel coding, the Polar code launched by Huawei and other companies is currently the channel coding

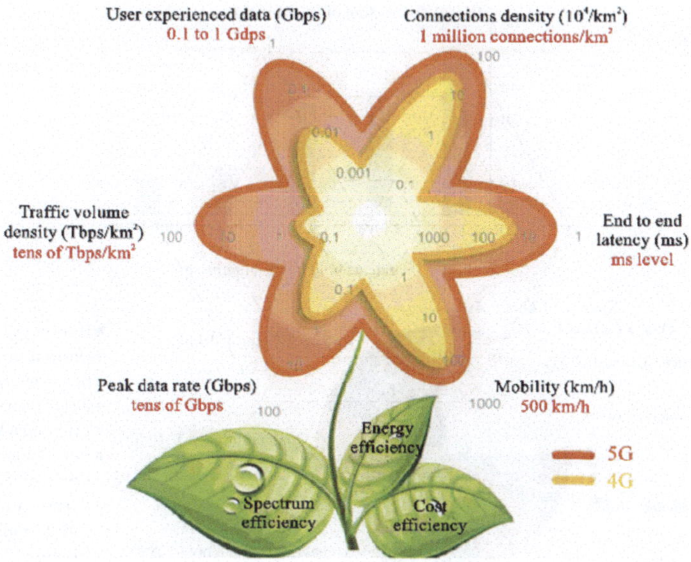

Fig. 3.8 "5G Flower" standard proposed by China. *Image Source* IMT-2020 (5G) Promotion Group, 5G Vision and Requirements White Paper, 2014

method that has been strictly proved to reach the Shannon limit, and it has determined the channel coding technology solution for the 5G eMBB (enhanced mobile broadband) scenario. In the core network field, China Mobile took the lead and jointly proposed service based architecture with 14 global operators and 12 network equipment vendors including Huawei, which was officially confirmed by 3GPP as the only infrastructure standard for the 5G core network.

Some core technologies of the 5G standard are also protected by patents, and the relevant patents are called 5G standard essential patents. The technical directions of 5G standard essential patents include wireless resource management, access technology, multi-carrier transmission, channel coding, core network, and next-generation access network. The current 5G communications standards R15, R16 and R17 have all completed the "frozen version", and the corresponding standard essential patent layout has basically become a foregone conclusion. The database search results of the German patent database company IPlytics showed that as of February 2022, the number of 5G standard essential patents owned by Chinese enterprises had ranked second among the relevant countries/regions in the world, accounting for 38.4% (see Fig. 3.9). As of December 2022, the number of 5G standard essential patents declared worldwide had exceeded 84,900, of which the top 10 enterprises contributed 75% of the total number of patent families; the number of effective global patent families of Chinese enterprise Huawei accounted for 14.6%, ranking first among global enterprises, and the number of effective global patent families of other Chinese enterprises such as ZTE, Datang, OPPO, and Xiaomi also ranked among the top ten in the world.

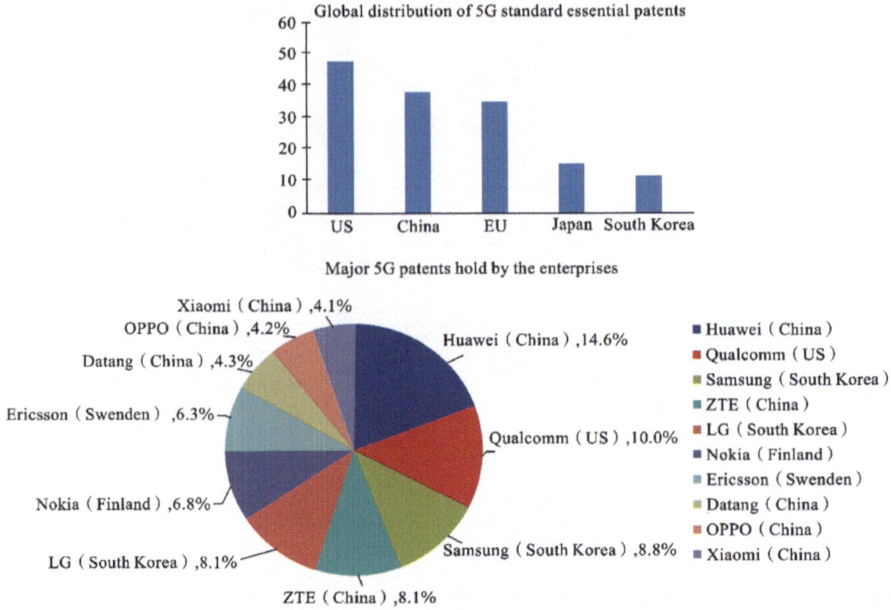

Fig. 3.9 Global distribution of 5G standard essential patents (February 2022). *Data Sources* China National Intellectual Property Administration (top), China Academy of Information and Communications Technology (bottom)

In terms of 5G commercial use, China launched a commercial 5G technology research and development test in 2016. In December 2018, the MIIT issued 5G system low-and medium-band test frequency use licenses to China Telecom, China Mobile, and China Unicom. In order to reduce the investment in 5G construction, under the leadership of the government, China Telecom and China Unicom achieved the co-construction and sharing of 5G network base stations through China Tower Corporation, sharing more than 300,000 5G base stations, and saving more than 60 billion yuan in construction investment. By the end of 2023, China had built and opened more than 3.377 million 5G base stations, achieving full coverage of cities at or above the prefecture level across the country, with 805 million 5G mobile phone users and 47% of 5G network access traffic. From January to December 2023, China's domestic 5G mobile phone shipments reached 240 million units, an annual increase of 11.9%, accounting for 82.8% of mobile phone shipments in the same period. China has become the country with the largest scale of 5G applications in the world.

With the advancement of commercial use, 5G, a new engine for the development of the digital economy, will not only open up new development space for the ICT industry, but will also be deeply integrated with the real economy. Commercial 5G professional modules have also been launched, which can support 5G 8K TVs, 5G industrial production lines, smart transportation and other industry applications. A related report from the China Academy of Information and Communications Technology predicted that in 2025, 5G will drive China's economic added value by about 1.1 trillion yuan, contributing 3.2% to the GDP growth of that year, and indirectly driving GDP by 2.1 trillion yuan; in 2030, 5G is expected to contribute more than 2.9 trillion yuan to China's economic added value, with a compound annual growth rate of 41% over the past 10 years.

3.4.3 Integrated Circuit Industry

Integrated circuits are known as the foundation of the ICT industry. The integrated circuit industry is a strategic industry that affects the comprehensive competitiveness of a country's economy, politics, and national defense. Its technical level and industrial scale have become an important sign to measure a country's industrial competitiveness and comprehensive national strength.

Integrated circuits include multiple fields, the most representative of which is the field of memory chips. Data showed that the global semiconductor market scale exceeded 180 billion USD in 2022, with memory chips accounting for about 35%. The memory chip market could be further divided into the memory (DRAM) market and the flash memory (NAND) market, of which the memory market scale accounted for about 61% of the global storage market scale, while the flash memory market scale accounted for about 36%. A few years ago, the international semiconductor memory field was mainly monopolized by several enterprises such as Samsung and Hynix in South Korea, Micron in the US, and Toshiba in Japan. China's annual chip imports amounted to nearly 2.8 trillion yuan, of which the import value of memory chips was about 700 billion yuan.

In order to solve the common technical problems of the integrated circuit industry, China launched the National Science and Technology Major Project "Very Large Scale Integrated Circuit Manufacturing Equipment and Complete Process" in 2008. The overall goal of this project was to carry out research on integrated circuit manufacturing equipment, complete process and material technology, and master core technologies. Since the implementation of this project, more than 200 enterprises and institutions and more than 20,000 scientific researchers have participated in technical research. Before 2008, the most advanced mass production process of integrated circuit manufacturing in China was 130 nm, and the process level of research and development was 90 nm; after the implementation of this project, by 2017, the mainstream process level had been improved by 5 generations, and the three generations of complete process of 55 nm, 40 nm and 28 nm had been successfully developed and mass-produced, and the research and development of 22 nm and 14 nm leading technologies had made breakthroughs and formed independent intellectual property rights. China has successfully developed more than 30 high-end equipment such as 14 nm etcher and thin film deposition, and hundreds of material products such as target materials and polishing liquids, with performance reaching international advanced levels.

In order to solve the financing support problem of the integrated circuit industry, the State Council of China issued the "National Outline of Promoting the Development of Integrated Circuit Industry" in June 2014. In the same year, the National Integrated Circuit Industry Investment Fund Co., Ltd. (hereinafter referred to as the "National Big Fund") was established. In 2019, the National Integrated Circuit Industry Investment Fund Phase II Co., Ltd. (hereinafter referred to as the "National Big Fund Phase II") was established. As of December 2021, the National Big Fund and the National Big Fund Phase II had completed investments in 83 integrated circuit entity companies, covering the backbone enterprises in the industrial chain, such as SMIC, Huahong Group, and China Resources Microelectronics in the wafer manufacturing industry, Yangtze Memory, Changxin Storage, and GigaDevice in the storage industry, Changdian Technology, Tongfu Microelectronics, and Huatian Technology in the packaging and testing industry, Guowei and Huada Jiutian in the electronic design automation tool industry, Unisoc, and Zhixin Microelectronics in the design industry, North Huachuang, and China Microelectronics in the equipment and parts industry, and Yak Technology and ANJI in the materials industry.

By taking the above measures, many problems of Chinese semiconductor enterprises in technology, capital and other aspects have been solved, and domestic enterprises have made breakthroughs in multiple semiconductor fields.

In terms of memory chips, Changxin Storage was jointly established in Hefei in May 2016 by the Hefei Municipal Government, GigaDevice, and the National Big Fund. It is China's first independent memory chip manufacturing project, with a total investment of over 200 billion yuan. Changxin Storage purchased the memory technology of the German company Qimonda and completed the original technology accumulation. Later, Changxin Storage invested 2.5 billion USD in research and development expenses for upgrading. In the first quarter of 2018, Changxin Storage completed the equipment installation of the first 12-inch (30.48 cm) wafer factory, and developed China's first 8 Gb DDR4 chip in the same year. In the third quarter of 2019, Changxin Storage successfully mass-produced DDR4 memory (see Fig. 3.10) and LPDDR4 memory using the 19-nm process technology, achieving a yield of 70% to 75%, and the maximum speed can reach 3200 Mbps, becoming the world's fourth manufacturer that can mass-produce memory chips below 20 nm. Changxin Storage had achieved a memory chip production capacity of 60,000 pieces/month in 2021, and the future goal was to achieve a production capacity of 300,000 pieces/month.

Fig. 3.10 DDR4 memory module launched by Changxin Storage. *Image Source* Changxin Storage

In terms of flash memory chips, Yangtze Memory was established in Wuhan in July 2016. Investors include Tsinghua Unigroup, Hubei Integrated Circuit Industry Investment Fund, China Development Fund, and National Big Fund. Yangtze Memory and Phison Semiconductor completed the accumulation of basic technologies in cooperation, and then jointly developed with the Institute of Microelectronics of the Chinese Academy of Sciences. In February 2017, the Institute of Microelectronics of the Chinese Academy of Sciences announced that the domestically produced 32-layer 3D NAND flash memory chip passed various tests, including those for electrical characteristics. In August 2018, Yangtze Memory released Xtacking technology at the Flash Memory World (FMW). This technology processes storage units and logic circuits separately on two wafers, and then connects the two wafers together through millions of metal channels, increasing the maximum access speed of flash memory chips to the level of DDR4 memory. In September 2019, Yangtze Memory created the world's first 64-layer TLC 3D NAND flash memory chip based on the Xtacking architecture and achieved mass production (see Fig. 3.11). In April 2020, Yangtze Memory launched the world's first 128-layer QLC 3D NAND flash memory chip, which has the largest storage capacity and the highest I/O interface transmission speed. At the end of 2021, Yangtze Memory reached a production capacity of 100,000 wafers per month. In November 2022, Yangtze Memory achieved mass production of 232-layer 3D NAND flash memory chips, becoming the first NAND storage manufacturer in the world to achieve mass production of NAND flash memory chips with more than 200 layers. The I/O speed of the NAND flash X3-9070 reached 2400 MT/s (million transfers per second), with a 50% performance improvement and a 25% reduction in power consumption.

After nearly six years, China's semiconductor storage enterprises represented by Changxin Storage and Yangtze Memory integrated the achievements of China's scientific and technological projects, actively forged ahead, aggressively innovated, and made pioneering breakthroughs in the storage field. According to statistics, the market share of Chinese manufacturers in the global memory chip market was still 0 in 2015, but by 2022, their combined market share in the global DRAM + NAND flash memory chip market was close to 5%. The rapid growth of these enterprises was

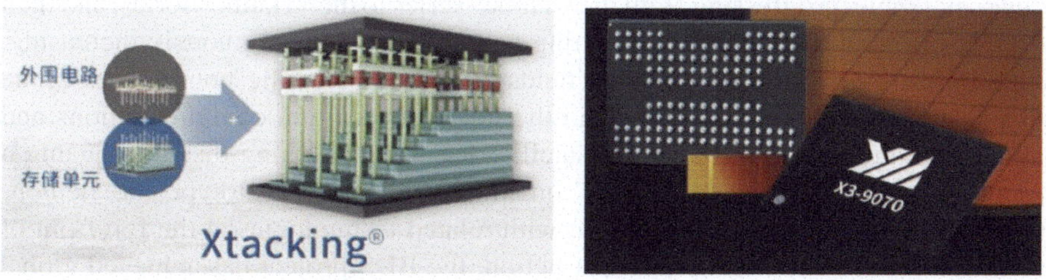

Fig. 3.11 Xtacking-based NAND flash memory launched by Yangtze Memory. *Image Source* Yangtze Memory

inseparable from the support of national policies, the sharing of scientific and technological achievements of state-owned scientific research institutes, and the backing of the National Big Fund.

3.5 Summary

At the beginning of the founding of the People's Republic of China, the foundation for ICT research and development was extremely weak. After more than seven decades of development, China has gradually established a complete system for scientific research and technological development. China has established a strategic scientific and technological force system, and institutions such as national key laboratories and NERCs are responsible for undertaking research in basic theories and key technologies. At the same time, China has established various types of industrial innovation bases, including HIDZs and innovative industrial clusters, to encourage enterprises to carry out technological research and innovation. Currently, China has achieved breakthrough progress in fields such as integrated circuits and new display technologies.

China has established comprehensive science and technology development planning systems. Through different levels of plans, such as the National Natural Science Foundation of China, National Science and Technology Major Project, and National Key Research and Development Programs of China, China guides domestic scientific research institutions and enterprises to conduct research in key fields. China has also deployed a number of national key projects for "Digital China" construction, covering areas such as digital infrastructure and the digital economy, and laying the foundation for the digital transformation and upgrading of various industries. China has established a multi-dimensional mechanism for R&D investment. The state's investment in basic theoretical research has been continuously increasing, and enterprises' investment in product development has also been gradually enhanced.

In terms of major scientific and technological projects, China has achieved a number of world-renowned achievements. The development of supercomputers has been remarkable, progressing from the "Yinhe" series to the "Tianhe" series, and then to the "Sunway Taihu-Light", computing power has been continuously increasing, with applications spanning multiple fields including scientific computing. China has made outstanding achievements in the fields of quantum communications and quantum computers. The "Micius" satellite has achieved long-distance quantum communications, and the "Jiuzhang" quantum computing prototype has demonstrated super-strong computing power, with related technologies at the forefront of the world. After three rounds of construction, the BDS project has achieved global networking and provided advanced civil global navigation services.

In terms of tackling key problems in information industry technologies, China has also made remarkable progress. The Ultra-HD video industry has established digital television standards, and the industrial ecosystem has gradually been improved, leading the development of 4 K Ultra-HD technology. The mobile communications industry has achieved a leap from following to leading. China has made breakthroughs in aspects such as 5G baseband chips and air interface technologies. China leads globally in the number of essential patents for the 5G standard of the new-generation mobile communications, and boasts the world's largest commercial scale. With the support of national special projects and funds, the IC industry has achieved major breakthroughs in the field of memory chips, gradually overcoming the technical bottlenecks.

China has achieved all-round development and progress in ICT innovation. The national education system established by China over the years has developed a well-structured engineer workforce for ICT industry. The strategic scientific research forces have provided a crucial support for solving theoretical problems in information science. The science and technology development planning systems established have formed an efficient scientific research organization model, and completed major scientific infrastructure projects such as supercomputers and quantum computers. Through measures such as establishing HIDZs, innovative industrial clusters, and industrial development funds, the Chinese government encourages enterprises to carry out technological innovation and has achieved breakthroughs in multiple industrial technologies.

Chapter 4
Digital Industrialization

Digital industrialization, that is, the industrialization of digital technology, refers to the transformation of digital technology into marketable goods and services, thereby forming industries with economic scale and market competitiveness. The content of digital industrialization involves aspects such as digital product manufacturing, digital product services, digital technology applications, and etc. Its scope includes traditional information industries and emerging digital application industries. For the convenience of readers, this chapter is organized according to the classification of traditional information industries (fundamental telecommunications industry, Internet industry, software and information technology services industry, and electronic information manufacturing industry) to present the development achievements China has made in recent years.

4.1 Fundamental Telecommunications Industry

From 2012 to 2021, China's information and communications technology (ICT) industry grasped the opportunities of global development, and quickly integrated into the international industrial division of labor system. The modern ICT business, represented by the telecommunications sector, rose rapidly. In around 10 years, the average growth rate of China's telecom business reached more than 20%, far exceeding the average GDP growth rate during the same period. Especially in 2017 and 2018, the growth rate of total telecom business was as high as 76.7% and 137.5% respectively. In 2021, the total business of China Telecom reached 1.7 trillion yuan.

W. Liu et al. (eds.), *Digital China*, Introduction to China's S&T Innovation,
https://doi.org/10.1007/978-981-95-6240-4_4

4.1.1 *Internet Infrastructure*

4.1.1.1 Optical Fiber Backbone Network

In 1994, China established several high-performance backbone networks covering all cities nationwide. After 2000, a hierarchical architecture of interconnected backbone networks emerged, with Beijing, Shanghai, and Guangzhou as the main backbone direct connection points, supplemented by exchange centers. In 2013, seven new backbone direct connection points were added in the city of Chengdu, Wuhan, Xi'an, Shenyang, Nanjing, Chongqing, and Zhengzhou. The activation of these new backbone direct connection points resulted in a performance enhancement of over 60%, significantly enhancing the backbone network intercommunication efficiency and forming a balanced regional structure.

The construction of the optical fiber backbone network made remarkable achievements. China has built the world's largest optical fiber broadband network, occupying about half of the global optical fiber market demand since 2014. By the end of 2021, the total length of optical fiber cable lines in China had reached 54.81 million kilometers, an increase of approximately 2.7 times compared with that of 2012. Concurrently, backbone transmission had evolved towards ultra-high-speed, high-capacity networks, fully establishing a national single-wave 100 Gbps backbone network; the transmission speed of the backbone network had entered the 200 Gbps era, with 400 Gbps transmission systems piloted in some cities. The trends of the total length of optical cable lines and the length of trunk lines in China from 2016 to June 2021 are shown in Fig. 4.1.

The comprehensive transmission capacity of the network has been continuously improved. As of June 2021, a total of 14 backbone direct connection points had

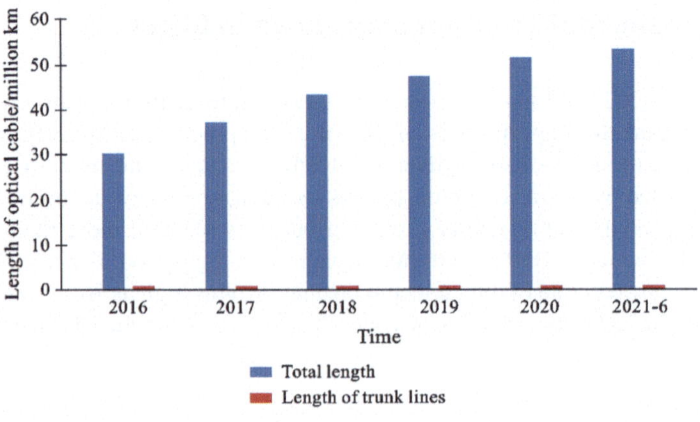

Fig. 4.1 The Trends of the total length of optical cable lines and the length of trunk lines in China (2016-June 2021). *Data Source* Ministry of Industry and Information Technology of the People's Republic of China

been built in China. Taking the National (Hangzhou) Novel Internet Exchange as an example, its total access bandwidth reaches 2 Tbps, the comprehensive, multi-level, and three-dimensional network interconnection architecture has been further optimized, and the performance of inter-network communications has continued to improve. As of the end of June 2021, China's internation Internet bandwidth reached 8.6 Tbps, representing a year-on-year increase of 31.6%.

The network infrastructure fully supports IPv6. China has built the world's largest IPv6 network infrastructure. As of December 2021, the number of IPv6 addresses in China had reached 60059 (/32), ranking first in the world; the number of active IPv6 users in China had reached 608 million, accounting for 60.11% of the total number of Internet users; IPv6 network traffic had grown rapidly, with mobile network IPv6 traffic accounting for 35.15% of the total mobile network traffic.

The international interconnection facilities have been further upgraded. The pace of the related construction such as submarine optical cables and cross-border terrestrial cables has been accelerated, and the interconnection level with countries jointly building the "Belt and Road" has been steadily improved. By the end of June 2021, China's Internet international import and export bandwidth had reached 8.6 Tbps, a year-on-year increase of 31.6%. By the end of 2020, China Telecom, China Mobile, and China Unicom had built more than 400 overseas Internet points of presence (POPs).

4.1.1.2 Optical Fiber Access Network

The broadband access network continues to be upgraded. In March 2021, the Ministry of Industry and Information Technology (MIIT) released the "Action Plan for the Coordinated Development of 'Dual-Gigabit' Internet (2021–2023)". The "dual-gigabit" Internet, represented by gigabit optical network and 5G network, can provide a single user with gigabit access capability of fixed and mobile networks. It has the characteristics of super-large bandwidth, ultra-low delay, and advanced reliability, and is an important component and carrying base of new infrastructure.

Gigabit optical fiber has covered one-third of China's households. By the end of June 2021, the total number of optical fiber to the home/office (FTTH/O) ports in China had reached 920 million, accounting for 93.5% of all broadband access ports, an increase of 1.4 percentage points from the same period last year. The optical fiber access capacity generally exceeds 100M, and is further upgraded to a speed of more than 1000M. As of June 2021, 36 provincial-level operators across China had released fiber-to-the-room (FTTR) packages, and the pilot program-basically covered the whole country. Home and enterprise ultra-high-speed LAN networking solutions were gradually promoted and applied. In December 2021, 29 cities in China became the first batch of gigabit cities, having built 7.86 million 10 Gigabit Passive Optical Network (10 G-PON) ports, with gigabit optical networks capable of covering over 300 million households. The trends of the number and proportion of FTTH/O ports in China from 2016 to June 2021 are shown in Fig. 4.2.

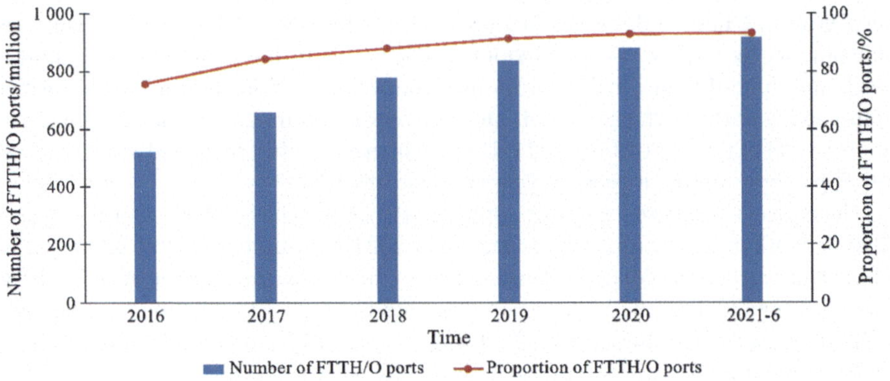

Fig. 4.2 The trends of the number and proportion of FTTH/O ports in China (2016–June 2021). *Data Source* Ministry of Industry and Information Technology of the People's Republic of China

The users' actual Internet access experience has been greatly improved. In 2012, China's average access network speed was only about 2 Mbps, ranking in the middle and lower levels among major countries in the world. In 2021, the average download speed of China's mobile network had reached 59.34 Mbps, and the average download speed of fixed networks had exceeded 62.55 Mbps, ranking among the top in the world; the average download speed of rural optical fibers had also exceeded 100 Mbps, and infrastructure construction had achieved improvements in both network coverage and network speed. According to the monitoring data released by the network speed test and statistics company Ookla, in September 2021, the average download speed of China's fixed broadband ranked 15th among 181 countries.

4.1.1.3 Mobile Communications Network

In the 10 years from 2012 to 2021, the mobile communications sector in China had successfully achieved several technological breakthroughs, progressing from "breakthrough in 3G", "synchronization in 4G" to "leading in 5G". In 2013, China realized the commercialization of 4G; in 2019, China realized the commercialization of 5G. In 2021, China's 5G patent applications exploded, and the number of mobile communications patents accounted for more than 30% of the world's total. China has become the world's primary producer of 5G patents and a standard-setting country, realizing the leadership of mobile communications technology.

In parallel with technological breakthroughs, the base station construction are advancing rapidly. By the end of 2019, China had built 8.41 million mobile communications base stations, with 5.44 million 4G base stations. By the end of 2021, over two years into the 5G era, China had 9.96 million mobile communications base stations, an 18.4% increase from 2019, with 5.9 million 4G base stations and 1.425 million 5G base stations; the 5G network had already covered all urban areas of prefecture-level cities, over 98% of county-level urban areas, and 80% of township-level urban areas; over 99% of the administrative villages had 4G coverage. Internationally, in the 5G era, China took the lead in 5G base station construction, with over 60% of the world's 5G base stations located in China, making it a global leader in 5G network construction. Figure 4.3 shows the trends of the number and proportion of 5G base stations in China from March 2020 to June 2021.

Mobile terminals are becoming increasingly widespread. According to the statistics of the MIIT and the Global System for Mobile Communications Association (GSMA), by 2021, the number of mobile users in China had reached 1.64 billion, with a penetration rate of 116.3 per one hundred people. The number of 5G mobile phone terminal connections in China had rapidly increased, exceeding 490 million. As 4G users gradually migrated to 5G, the number of 4G users in 2021 decreased by 12% compared with 2020, totaling approximately 1.13 billion, with a penetration rate of 68.8%. In 2021, China's mobile data traffic continued to grow, with the average monthly mobile Internet access traffic (discharge of usage, DOU) per user in October being 14.32GB, about 23.7% higher than the global average.

Mobile data tariffs continue to decline. In the first half of 2021, the average mobile data tariff in China fell to 3.2 yuan/GB, a 24.6% decrease from the previous year; and the average monthly mobile data usage per user was 12.6GB, a year-on-year increase of 32.6%. The substantial growth in mobile data consumption led to a year-on-year increase of 4.8% in average revenue per user (ARPU), reaching 49.9 yuan. From an international comparison perspective, according to the GSMA statistics, in the fourth quarter of 2020, China's mobile communications tariffs were among

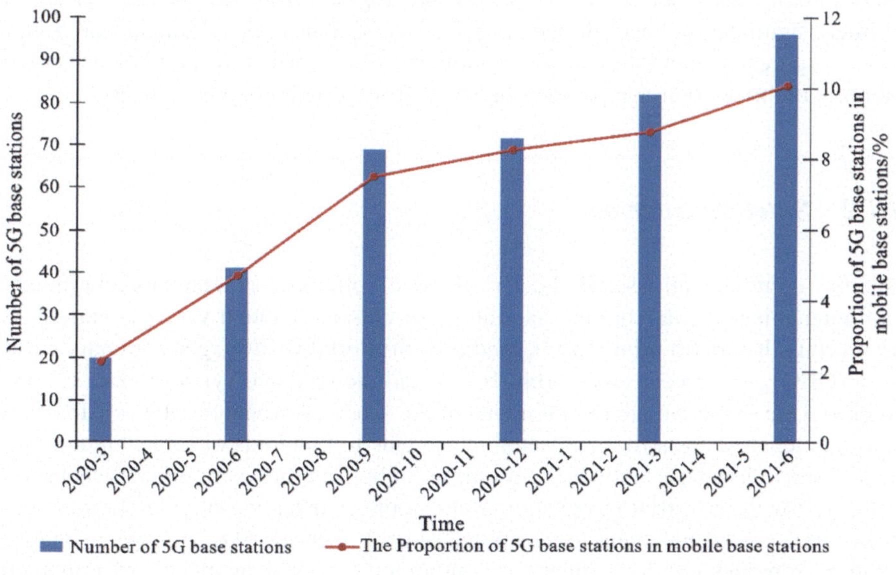

Fig. 4.3 The trends of the number and proportion of 5G base stations in China (March 2020–June 2021). *Data Source* Ministry of Industry and Information Technology of the People's Republic of China

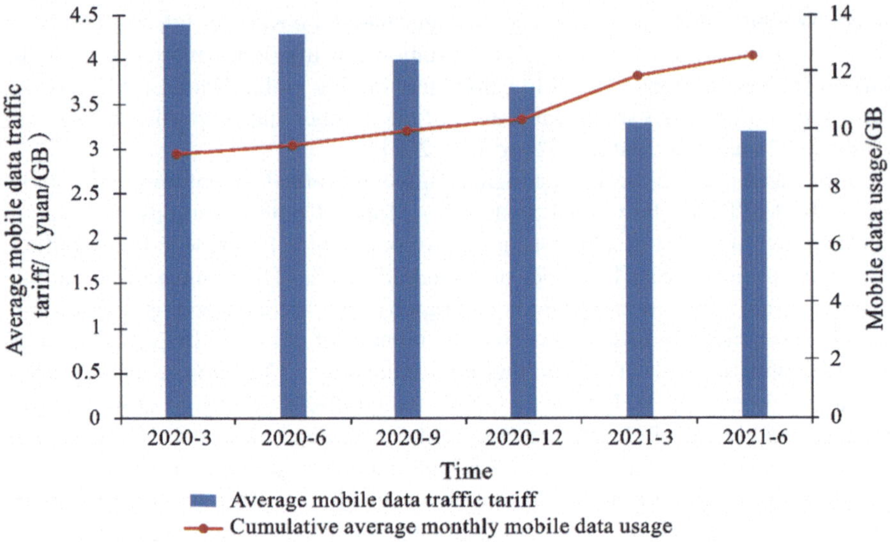

Fig. 4.4 The average mobile data traffic tariff and cumulative average monthly mobile data usage per user in China (March 2020-June 2021). *Data Source* Ministry of Industry and Information Technology of the People's Republic of China

the lowest globally, with the average monthly expenditure of mobile communication users ranking 93rd out of 237 countries and regions from low to high in terms of price, significantly lower than that of the United States (US), Canada, and South Korea. Figure 4.4 shows the average mobile data traffic tariff and cumulative average monthly mobile data usage per user in China from March 2020 to June 2021.

4.1.2 Satellite Internet

Satellite communications will be an effective solution for achieving ubiquitous communications in the future. Satellites are classified into five types based on their orbits: low earth orbit (LEO), medium earth orbit (MEO), geostationary earth orbit (GEO), sun-synchronous orbit (SSO), and inclined geo-synchronization orbit (IGSO). Due to the unique requirements of the space environment, the construction and maintenance costs for each satellite are tens to hundreds of times higher than those for wireless base stations, resulting in higher satellite communications tariffs compared with terrestrial mobile communications. Currently, satellite communications serves as a supplement to traditional optical fiber and 5G coverage, primarily targeting remote areas that optical fiber cannot reach, as well as specific regions such as oceans, deserts, and polar regions. This approach shifts network coverage focus from population centers to territorial expanses.

In recent years, LEO broadband communications satellites have garnered significant industry attention. These satellites offer lower launch costs and shorter round-trip times, making them a primary means for achieving satellite Internet and a key consideration for next-generation 6G mobile communications standards. LEO broadband communications satellites use the inter-satellite link technology to create a space backbone network via laser or microwave links. Satellite communications terminals can transmit data via orbital satellite constellations, through the space backbone network or directly through gateway stations, facilitating international Internet communications. SpaceX's Starlink project has launched numerous LEO broadband communications satellites, and by August 2021, it had covered all regions of the Earth except the Antarctic and Arctic, and related services have been deployed in the US, Canada, and Europe, with a planned deployment of 12,000 satellites by 2027.

In terms of high-throughput satellites, China has established a satellite network composed of three high-throughput satellites. In April 2017, China launched its first high-throughput satellite, Zhongxing-16, with a Ka-band multi-beam broadband communications capacity of 20 Gbps, covering most parts of China. By July 2020, it passed commercial flight network tests, achieving an access speed of over 150 Mbps. In November 2022, China launched the Zhongxing-19 high-throughput satellite to cover the Pacific region. In February 2023, China launched the Zhongxing-26 high-throughput satellite, covering China and its surrounding areas. At present, the Zhongxing-26 satellite is China's most advanced civil and commercial communications satellite, boasting a communications capacity of over 100 Gbps, equipped with 94 user beams and 11 gateway beams, and offering a terminal communications speed up to 450 Mbps. Together, these three high-throughput satellites provide high-speed network communications services and satellite Internet access to users.

In terms of GEO satellite mobile communications systems, China has launched three "Tiantong-1" satellites. In August 2016, China launched the Tiantong-1 01 satellite, which was the first satellite of China's satellite mobile communications system; subsequently, in November 2020 and January 2021, China launched the Tiantong-1 02 and Tiantong-1 03 satellites, respectively. These satellites, together with China's terrestrial mobile communications system, have formed an integrated space-ground mobile communications network, providing all-weather, all-time, stable, and reliable mobile communications services for users in China, neighboring Asian regions, and parts of the Pacific and Indian Oceans. In August 2023, Huawei released the smart phone of Mate 60 Pro, which supports access to the Tiantong-1 satellite mobile communications system. It is the world's first regular commercial mobile phone that supports satellite mobile communications, allowing users to make and receive satellite calls in areas where mobile communications networks are difficult to cover, such as mountains, deserts, and islands.

In terms of LEO satellite communications networks, China's two major aerospace groups, China Aerospace Science and Technology Corporation Limited, and China Aerospace Science and Industry Corporation, have initiated their own LEO network projects: the "Hongyan Satellite Constellation" planning to launch 300 small satellites, and the "Hongyun Project" planning to launch 156 satellites. Both projects completed their first satellite launches in 2018. In April 2020, satellite Internet was designated as one of China's "New infrastructure", marking 2020 as the inaugural year for satellite Internet construction in China. The "Outline of the 14th Five-Year Plan for National Economic and Social Development and the Long-Range Objectives Through the Year 2035 of the People's Republic of China" released in March 2021, called for the construction of high-speed, ubiquitous, integrated, interconnected secure, and efficient space-ground information infrastructure. In April 2021, China Satellite Network Group Co., Ltd (China SatNet) was established to specialize in the design, construction, and operation of satellite Internet. China has submitted network application to the International Telecommunication Union (ITU) for LEO constellation orbits and frequencies, totaling 12,992 satellites in orbits ranging from 590 to 1145 km above the Earth. China SatNet plans to launch about 10% of these satellites within five years, supporting the deployment of the 6G mobile communications network by 2035.

4.1.3 Internet of Things

The Internet of Things (IoT) is a digital information infrastructure that uses a collaborative multi-network cellular mobile communications network to achieve ubiquitous connectivity and the interconnection of all things. International communications standardization organizations have successively completed various IoT-related technical standards, ranging from narrowband and low-speed to high-speed and low-latency. Among them, narrowband IoT (NB-IoT) meets most low-speed IoT requirements, 4G Cat.1 (a 4G network with speed category 1, including LTE-Cat.1) meets medium IoT and voice requirements, 5G RedCap meets mediums to high-speed requirements, and 5G NR meets high-speed and low-latency IoT requirements.

In scenarios with wide coverage, low sensitivity to latency, and low data rates, such as intelligent metering, tracking and positioning, intelligent smoke detection, and smart parking, NB-IoT technology is suitable. For scenarios where requirements for data rates are moderate but power consumption and transmission stability is critical, such as Quick Response (QR) code payments and charging stations, 4G Cat.1 technology is ideal. For applications requiring higher data rates along with specific constraints on size, cost, and power consumption, such as video surveillance, industrial sensing, medical monitoring, and high-end wearable devices, 5G RedCap technology is appropriate. In scenarios demanding ultra-high bandwidth, low latency, and high mobility, such as industrial IoT, smart healthcare, intelligent transportation, and entertainment, 5G NR technology is the best fit.

China places significant emphasis on mobile IoT, continuously driving its development. In May 2020, the "Notice of the General Office of the Ministry of Industry and Information Technology on Deeply Advancing the Comprehensive Development of Mobile Internet of Things" was issued, which proposed migrating 2G/3G IoT services and creating an integrated mobile IoT ecosystem that facilitates coordinated development of NB-IoT, 4G, and 5G. Chinese telecom operators are promoting the adoption of NB-IoT terminals through subsidies. According to the 2021 "China

Narrowband IoT (NB-IoT) Industry Market Research" report, since 2017, China Mobile had allocated 1 billion yuan in subsidies for NB-IoT modules, with the maximum subsidy rate for a single module reaching up to 80%; China Telecom increased its subsidy rate for a single module to 50%, with a maximum subsidy of 30 yuan. The price of NB-IoT modules had fallen to around 15 yuan, nearly matching the cost of 2G IoT modules. NB-IoT had achieved millions of connections across various industries, and the large-scale implementation helps narrow the gap between revenue and connectivity.

China has established the world's largest mobile IoT, with 1.845 billion connections, representing over 70% of the global total. This makes China the first major economy to achieve "more connected devices than connected people". NB-IoT has formed millions to tens of millions of connections in sectors such as water management, gas, firefighting, tracking and positioning, door locks, and electric vehicle theft prevention. Mobile IoT has seen large-scale applications in digital city construction, intelligent manufacturing, smart transportation, mobile payments, and other fields, profoundly transforming people's lifestyle.

4.1.4 Internet of Vehicles

The Internet of Vehicles (IoV) uses new-generation ICT to achieve comprehensive network connections among vehicles, within vehicles, between vehicles and roads, between vehicles and people, and between vehicles and service platforms. IoV can enhance vehicle intelligence and autonomous driving capabilities, fostering new business models for automotive and transportation services. Regarding vehicle-to-vehicle networking standards and technologies, the 3GPP began the standardization of long term evolution-vehicle to everything (LTE-V2X) in early 2015, with China playing a leading role in international standardization efforts.

China has developed a complete industry chain based on LTE-V2X technology, with chips, modules, on-board units (OBUs), and roadside units (RSUs) that are mature and have undergone extensive testing, ready for commercial deployment. Regarding IoV standards and technologies based on 5G cellular networks, the 5G Automotive Association (5GAA) and Europe's 5GCroCo project conducted research on remote control driving scenarios; in 2019, China's IMT-2020 (5G) Promotion Group supported the establishment of the first testbeds integrating mobile edge computing and cellar-vehicle to everything (C-V2X).

In October 2018, the MIIT became the first globally to release the frequence regulation on IoV, titled "Interim Provisions on Managing the 5905–5925 MHz Frequency Band for Direct Communications of the Internet of Vehicles (Intelligent Connected Vehicles)", designating this frequency band for LTE-V2X based IoV direct communications. In February 2020, the National Development and Reform Commission (NDRC) and 10 other ministries issued the "Intelligent Vehicle Innovation Development Strategy", aiming for regional coverage of in-vehicle wireless communications networks (such as LTE-V2X) by 2025 and gradual application of 5G-V2X in some cities and highways, along with full coverage of high-precision space–time benchmark service networks. In March 2020, the MIIT issued the "Notice on Promoting the Accelerated Development of 5G", advocating for the coordinated development of "5G + IoV". In November 2020, the State Council issued the "New Energy Vehicle Industry Development Plan (2021–2035)", which further emphasized the deep integration of new energy vehicles with energy, transportation, and information communications sectors, and the coordinated advancement of intelligent road network infrastructure.

China has rapidly implemented new IoV infrastructure, yielding initial results. From 2019 to 2023, the MIIT successively approved seven national IoV pilot zones in Jiangsu (Wuxi), Tianjin (Xiqing), Hunan (Changsha), Chongqing (Liangjiang New Area), Hubei (Xiangyang), Zhejiang (Deqing), and Guangxi (Liuzhou). These zones actively promoted the construction of IoV infrastructure, interoperability testing, and large-scale pilot demonstrations.

In 2021, the Ministry of Housing and Urban–Rural Development (MHURD) and the MIIT designated 16 cities as pilot sites for the collaborative development of smart city infrastructure and intelligent connected vehicles. These cities are allowed to conduct demonstration applications of various levels of intelligent connected vehicles in specific scenarios. By October 2021, more than 20 cities and several highways in China had deployed over 4000 RSUs supporting direct communications, with over 3500 on urban roads and more than 500 on highways, covering a total road length of more than 3500 km.

4.1.5 Computing Infrastructure

In the digital economy era, computing power is emerging as a new form of productivity, providing essential support for the digital transformation of numerous industries. Data centers, as the physical hosts of computing power, are crucial infrastructure for digital development. With the widespread adoption of the Internet and ICT applications in China, the scale of data centers and servers has rapidly expanded. In 2021, the number of data center racks in use in China reached 5.2 million, with large and above data center racks accounting for 4.2 million, about 80% of the total. Data centers housed 19 million servers, with a storage capacity of 800EB. Figure 4.5 shows the scale of data center racks in use in China from 2017 to 2023.

However, the high energy consumption of large data centers is becoming a significant issue. For instance, in 2020, China had over 3 million data center racks in use, consuming more than 200 billion kilowatt-hours annually (based on a standard rack power density of 2.5 kilowatts). This consumption is nearly equivalent to the output of two Three Gorges hydropower station[1] and accounts for 2.7% of the country's total electricity use. Therefore, the energy conservation and consumption

[1] The Three Gorges hydropower station is the largest hydroelectric power station in the world and also the largest engineering project ever constructed in China. It officially started construction in

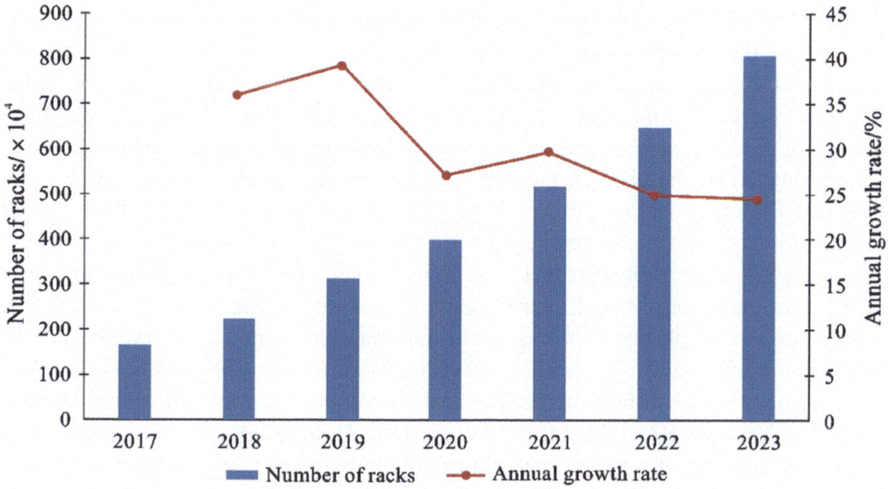

Fig. 4.5 The scale of data center racks in use in China (2017–2023). *Data Source* Ministry of Industry and Information Technology of the People's Republic of China

reduction of data centers have become the focus of attention for data center enterprises. Chinese enterprises have carried out active explorations. Take Alibaba Cloud as an example, it is the cloud service arm of Alibaba Group, has established five super data centers in Hangzhou, Nantong, Zhangbei, Ulanqab, and Heyuan in China. These data centers utilizes proprietary technologies to reduce energy consumption. For example, the Hangzhou data center uses the world's largest immersion liquid cooling server cluster, the Ulanqab data center relies on renewable sources of energy such as wind and solar power with natural wind cooling, and the Heyuan data center uses deep lake water for natural cooling.

From a broader geographical perspective, the contradiction between the computing power demands and energy supplies in various regions of China is becoming increasingly prominent. On the one hand, the economically developed eastern coastal regions of China have strong demands for computing power, but face a large electricity gap. On the other hand, the western regions of China are rich in various types of clean power generation resources, but have relatively lagging economic development with few demands for computing power. By building national systems for big data centers, a new type of computing power network system that integrates data centers, cloud computing, and big data can be constructed, and the computing power demand in eastern China can be guided to western China in an orderly manner. The green energy in the west can be consumed locally, solving long-standing problems such as high energy consumption of data centers and low utilization efficiency of data resources.

1994 and was completed in 2009. In 2023, it has completed a power generation of 802.71 billion kilowatt hours.

The "Outline of the 14th Five-Year Plan for National Economic and Social Development and the Long-Range Objectives Through the Year 2035 of the People's Republic of China" explicitly called for accelerating the construction of a national integrated big data center system, enhancing the unified and intelligent scheduling of computing power, and building several national hub nodes and big data center clusters. In May 2021, the NDRC, the MIIT, and other units jointly issued the "Implementation Plan for the Computing Hub of the National Integrated Big Data Center Collaborative Innovation System", proposing to carry out the "Eastern Data, Western Computing" demonstration project. It is an integrated layout super project adjusting and allocating national computing resources from a national perspective. The project included eight computing hub nodes, ten national data center clusters, and an all-optical interconnected high-speed network between these nodes and clusters. Data center clusters would be built in the four major economic regions of Beijing-Tianjin-Hebei, the Yangtze River Delta, the Guangdong-Hong Kong-Macao Greater Bay Area, and Chengdu-Chongqing, and computing hub nodes would be built in the four western regions of Inner Mongolia, Guizhou, Gansu, and Ningxia. The latter would take over the computing power needs of the former, and the computing power would be integrated and dispatched within the national system to realize western computing for eastern data.

The "Eastern Data, Western Computing" project can not only alleviate the pressure of China's rapidly growing computing power demand every year, but also alleviate the problem of unbalanced economic development between eastern and western China. The annual investment volume of this project will reach hundreds of billions of yuan, and the driving effect on related industries will reach 1:8. The data center industry chain includes traditional civil engineering projects as well as information technology equipment manufacturing, information and communications, basic software, green energy supply and other industries. It is long, covers a wide range of categories, and has a strong driving effect. The construction of computing power hubs and data center clusters will effectively drive upstream and downstream investment in industries in western China and promote social and economic growth.

4.2 Internet Industry

China's Internet industry has grown from scratch, from weak to strong, and now China has become a powerful Internet country in the world. China's top Internet enterprises include Alibaba, Tencent, Ant Douyin, Pinduoduo, and JD.com. (see Table 4.1) As of July 2022, the number of telecommunications business licensed enterprises across the country had surpassed 130,000, 4.9 times that of 2012, and the market players were more diversified; 10 enterprises ranked among the top 30 global Internet enterprises by market value, and their core competitiveness continues to improve.

In the domains of e-commerce, social media, entertainment media, and finance, the applications produced by China are globally dominant. In e-commerce, China's

Table 4.1 Representative Internet enterprises in China in 2022

Abbreviation	Full names	Business and brands	Locations
Tencent	Shenzhen Tencent Computer Systems Co., Ltd.	WeChat, QQ, Tencent Cloud	Guangdong
Alibaba	Alibaba (China) Co., Ltd.	Taobao, Alibaba Cloud, Youku, ele.me	Zhejiang
Meituan	Beijing Sankuai Online Technology Co., Ltd.	Meituan, Dianping, Meituan Waimai	Beijing
Ant Technology	Ant Technology Co., Ltd.	Alipay, AntChain, OceanBase Database	Zhejiang
Douyin	Beijing Douyin Information Service Co., Ltd.	TikTok, Today's Headlines, XiguaVideo	Beijing
JD. com	JD. com, Inc.	JD. com, JD Logistics, JD Technology	Beijing
Baidu	Baidu, Online Network Technology(Beijing)Co., Ltd.	Baidu Search, Baidu Cloud, BaiduApollo	Beijing
Pinduoduo	Shanghai Xunmeng Information Technology Co., Ltd.	Pinduoduo	Shanghai
Kuaishou	Beijing Kuaishou Technology Co., Ltd.	Kuaishou, Kuaishou Express, AC-Fun	Beijing
Trip	Trip. com Group Limited	Trip. com, Qunar, Skyscanner	Shanghai

Data Source Internet Society of China

transaction value and market scale have consistently been the largest in the world. In social media, WeChat and QQ boast nearly 1 billion monthly active users. In entertainment media, Chinese-developed apps like TikTok have become global sensations. In finance, third-party payment tools such as Alipay and WeChat led the world in application scale.

In the first half of 2022, various personal Internet applications in China continued to grow. Short videos experienced the most significant increase, with 28.05 million new users added since December 2021, a growth rate of 3.0%, pushing the usage rate of online videos to 94.6%. Instant messaging retained the highest user base, with 20.42 million new users since December 2021, reaching a usage rate of 97.7%. Online news and live streaming saw user increases of 16.98 million and 12.90 million with growth rates of 2.2% and 1.8%, respectively. Table 4.2 shows the user scale and usage rates of various Internet applications in China for the first half of 2022.

Table 4.2 User scale and usage rates of Internet applications in China, first half of 2022

Abbreviation	Full names	Business and brands	Locations
Tencent	Shenzhen Tencent Computer Systems Co., Ltd.	WeChat, QQ, Tencent Cloud	Guangdong
Alibaba	Alibaba (China) Co., Ltd.	Taobao, Alibaba Cloud, Youku, ele.me	Zhejiang
Meituan	Beijing Sankuai Online Technology Co., Ltd.	Meituan, Dianping, Meituan Waimai	Beijing
Ant Technology	Ant Technology Co., Ltd.	Alipay, AntChain, OceanBase Database	Zhejiang
Douyin	Beijing Douyin Information Service Co., Ltd.	TikTok, Today's Headlines, XiguaVideo	Beijing
JD. com	JD. com, Inc.	JD. com, JD Logistics, JD Technology	Beijing
Baidu	Baidu, Online Network Technology(Beijing)Co., Ltd.	Baidu Search, Baidu Cloud, BaiduApollo	Beijing
Pinduoduo	Shanghai Xunmeng Information Technology Co., Ltd.	Pinduoduo	Shanghai
Kuaishou	Beijing Kuaishou Technology Co., Ltd.	Kuaishou, Kuaishou Express, AC-Fun	Beijing
Trip	Trip. com Group Limited	Trip. com, Qunar, Skyscanner	Shanghai

4.2.1 Instant Messaging

From 1994 to 2009, the primary means of accessing the Internet was through personal computer (PC) terminals. In December 2009, the proportion of mobile phone Internet users in China surpassed 50% for the first time. In June 2012, the number of Internet users accessing the Internet via mobile phones reached 388 million, making mobile phones the most common Internet access terminal in China. Since 2010, with the rapid development of mobile Internet, the proportion of mobile phone Internet users among all Chinese Internet users rose from 66.2% in December 2010 to 99.6% in June 2022. Figure 4.6 shows the user scale of various Internet terminals in China from June 2002 to June 2022.

Instant messaging is the most frequently used Internet application among Chinese Internet users, with popular apps including WeChat and QQ. As of June 2022, the number of instant messaging users in China had reached 1.027 billion, an increase of 20.42 million since December 2021, accounting for 97.7% of all Internet users. Mobile Internet traffic grew rapidly, reaching 124.1 billion GB in the first half of 2022, a year-on-year increase of 20.2%.

WeChat has become an essential instant messaging tool for Chinese Internet users. Different from traditional instant messaging apps, WeChat integrates public accounts and mini-programs, allowing seamless access to third-party apps and creating a vast

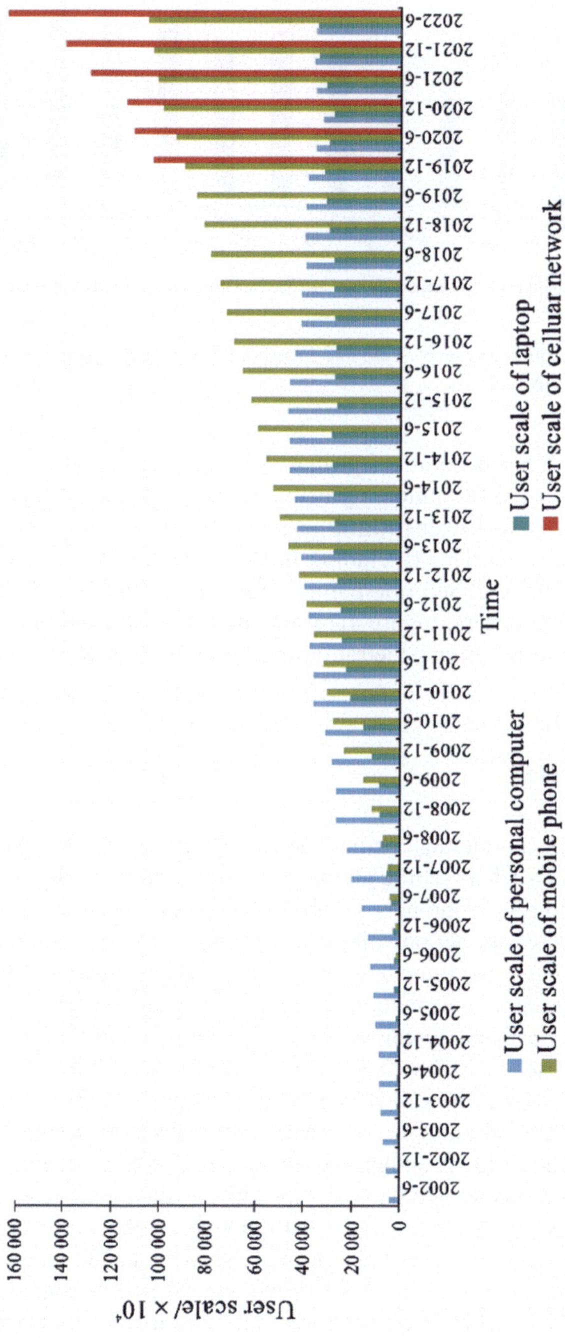

Fig. 4.6 The user scale of various Internet terminals in China (June 2002–June 2022). *Data Source* China Internet Network Information Center

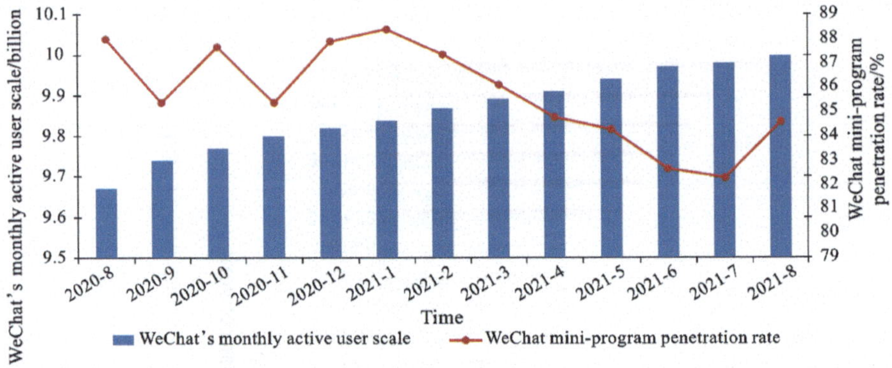

Fig. 4.7 WeChat's monthly active user scale and WeChat mini-program penetration rate (2020–2021). *Data Source* QuestMobile

software ecosystem. It has become the primary gateway for users to various mobile apps, maintaining over a billion monthly active users. In August 2021, the popular WeChat mini-programs (with over 100 million monthly active users) included various categories, such as city services (including health codes), utility payments, office documents, online shopping, and ride-hailing, covering essential daily activities of general Chinese Internet users. Figure 4.7 illustrates WeChat's monthly active user scale and WeChat mini-program penetration rate from 2020 to 2021.

4.2.2 E-commerce

E-commerce involves conducting business activities through the Internet and other information networks, such as selling goods or providing services. It is a crucial part of both the digital and real economies. China has long held the number one position globally in e-commerce transaction volume and market scale. E-commerce in China has become deeply integrated into all aspects of production and daily life, playing a pivotal role in the digital transformation of the economy and society.

According to the National Bureau of Statistics (NBS), the national e-commerce transaction volume reached 42.3 trillion yuan in 2021, with an annual growth rate of 19.6%. By transaction type, the goods transaction volume of 2021 was 31.3 trillion yuan, with an annual growth rate of 16.6%, while the transaction volume of services reached 11 trillion yuan, with an annual growth rate of 28.9%. By region, the eastern region's e-commerce volume of 2021 was 27.4 trillion yuan, with an annual growth rate of 18.2%; the central region's was 7.1 trillion yuan, with an annual growth rate of 22.6%; the western region's was 6.5 trillion yuan, with an annual growth rate of 24.2%; the northeastern region's was 1.3 trillion yuan, with an annual growth rate of 11.8%. Figure 4.8 shows China's e-commerce transaction value and annual growth rate from 2011 to 2021.

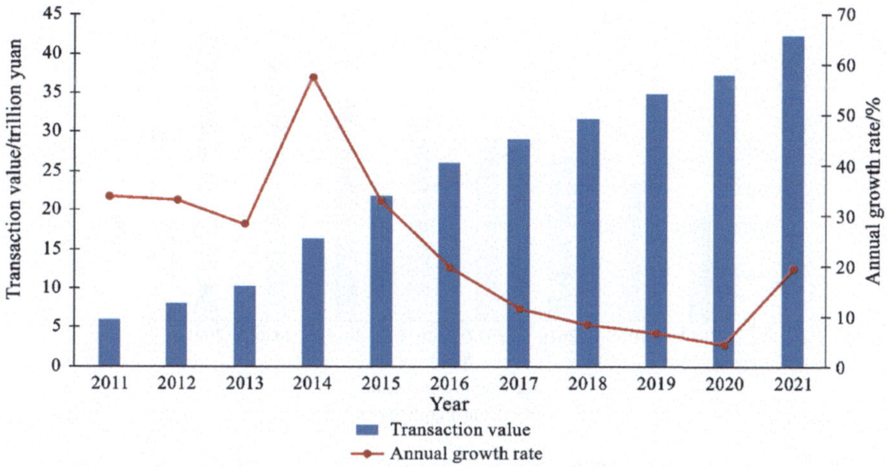

Fig. 4.8 China's e-commerce transaction value and annual growth rate (2011–2021). *Data Source* National Bureau of Statistics

Various new forms of e-commerce such as live streaming and community group buying have developed rapidly, accelerating the integration of e-commerce with the real economy and attracting more people to the e-commerce industry. According to the National Engineering Laboratory for E-commerce Transaction Technology, in 2021, the number of e-commerce practitioners in China reached 67.278 million, representing an 11.8% year-on-year increase. Among them, 41.2632 million were directly employed or started businesses in e-commerce, while 26.0148 million were in information technology, related services, and supporting industries driven by e-commerce. Figure 4.9 shows China's e-commerce employment scale and annual growth rate from 2014 to 2021.

4.2.2.1 Industrial E-commerce

Industrial e-commerce involves transactions between enterprises through third-party and self-operated B2B platforms. This includes bulk e-commerce for commodity trading, industrial products e-commerce for maintenance, repair and operations (MRO) and bill of material (BOM) services, wholesale e-commerce for consumer goods, enterprise procurement e-commerce for office supplies and business services, and related service providers. Key players in the industrial e-commerce in China include Alibaba, Beijing United Information Technology Co., Ltd., Netsun, HC Group, Focus Technology, Wangku Group, and Yiwu Purchase. Statistics showed that the scale of China's industrial e-commerce market grew from 6.25 trillion yuan in 2012 to 25.19 trillion yuan in 2021, a 3.03-fold increase over 10 years. Figure 4.10 illustrates the market scale and annual growth rate of China's industrial e-commerce from 2015 to 2023.

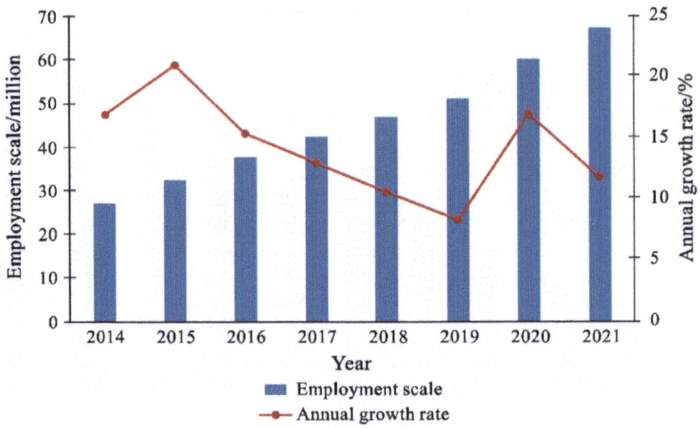

Fig. 4.9 China's e-commerce employment scale and annual growth rate (2014–2021). *Data Source* National Engineering Laboratory for E-commerce Transaction Technology

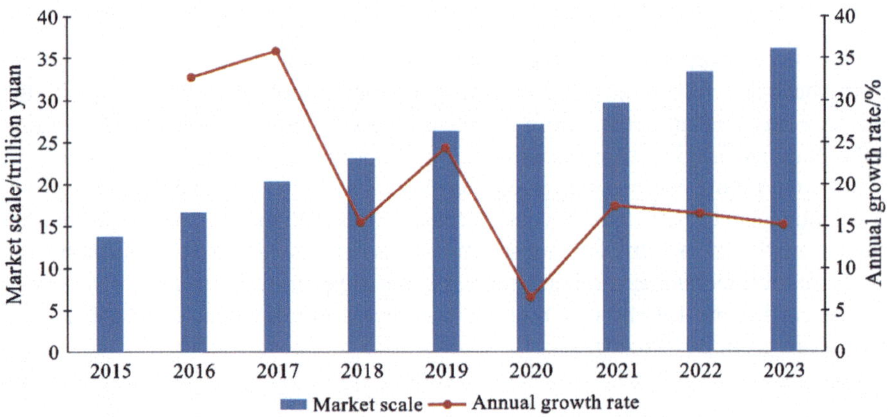

Fig. 4.10 The market scale and annual growth rate of China's industrial e-commerce (2015–2023). *Data Source* www.chyxx.com

4.2.2.2 Online Retail

According to the China Internet Network Information Center (CNNIC), as of December 2021, there had been 842 million online shopping users in China, accounting for 81.6% of all Internet users. Data from the NBS showed that in 2021, China's online retail value reached 13.09 trillion yuan, with an annual growth rate of 14.1%, maintaining the position of the world's largest online retail market for nine years. Online retail value of physical goods amounted to 10.8 trillion yuan, with an annual growth rate of 12.0%, representing 24.5% of total retail sales of consumer goods. In online retail value of physical goods, food, clothing, and daily necessities

Fig. 4.11 China's online retail value and annual growth rate (2011–2021). *Data Source* National Bureau of Statistics

grew by 17.8%, 8.3%, and 12.5%, respectively. In terms of categories of goods, the top three were apparel, daily necessities, and household appliances and audio-visual equipment, accounting for 22.94%, 15.23%, and 10.43% of physical goods online retail value, respectively. Figure 4.11 shows China's online retail value and annual growth rate from 2011 to 2021.

The "Double 11" shopping festival is a hallmark of China's online retail market. In November 2009, Tmall leveraged "Singles' Day" to establish a special shopping day. In 2019, Tmall's "Double 11" shopping festival set a sales record of 268.4 billion yuan, five times the transaction value of the Western "Black Friday" shopping festival that year.

Nowadays, the "Double 11" shopping festival is not only a major sales event for China's leading e-commerce platforms (Taobao, Tmall, JD.com,etc.), but also an annual global festival. Many international brands actively participate through national pavilions, overseas shopping, and direct brand sales. Data shows that from 2009 to 2021, the total transaction value of the "Double 11" shopping festival on Chinese e-commerce platforms increased annually, rising from 52 million yuan in 2009 to 965.12 billion yuan in 2021. Figure 4.12 shows the transaction value and annual growth rate of China's e-commerce platforms during the "Double 11" shopping festival from 2013 to 2021.

As more and more Internet platforms enter the e-commerce space, online shoppers are increasingly diversifying their consumption channels from traditional platforms like Taobao and JD.com to short video platforms, community group buying, and social media. Estimates from January to June 2022 indicate that 27.3% of online shoppers only used traditional e-commerce platforms, while 49.7% shopped on short video live streaming platforms, 37.2% on fresh food e-commerce, 32.4% on community group buying, and 19.6% on WeChat.

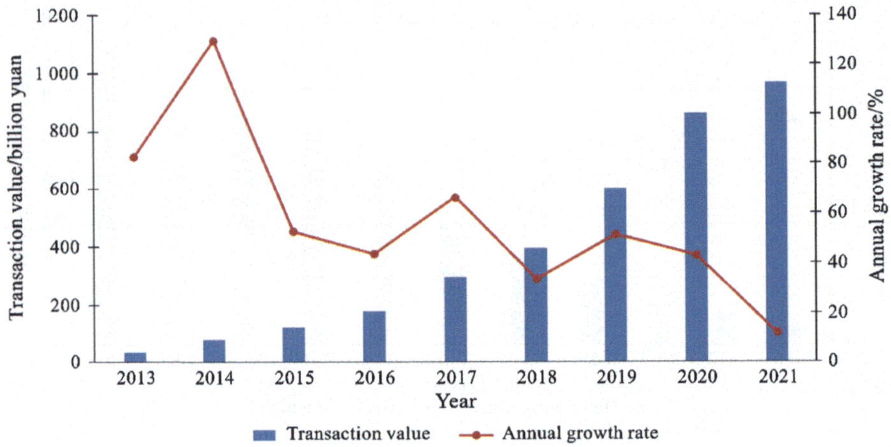

Fig. 4.12 China's e-commerce platform transaction value and annual growth rate during the "Double 11" shopping festival (2013–2021). *Data Source* data.iimedia.cn

4.2.2.3 Express Logistics

Online shopping has become a key consumption channel for Chinese residents, spurring rapid growth in related markets. In 2021, China's express delivery industry saw significant increases in both business volume and operating revenue. Specifically, according to the State Post Bureau of China, China's total business volume of express service enterprises reached 108.3 billion pieces, with an annual growth rate of 29.9%, while the total operating revenue hit 1033.23 billion yuan, with an annual growth rate of 17.5%. Intracity deliveries totaled 14.11 billion pieces, with an annual growth rate of 16.0%; intercity deliveries totaled 92.08 billion pieces, with an annual growth rate of 32.8%. Figure 4.13 shows China's express service enterprises business volume and annual growth rate from 2011 to 2021.

Various automated and unmanned systems have been integrated into logistics processes. In November 2016, the Nanjing Suning Super Cloud Warehouse, covering approximately 200,000 square meters, became Asia's largest intelligent logistics base. By 2019, it had been upgraded to a "next-generation unmanned warehouse", featuring various technical devices such as unmanned vehicles, fully automated transport robots, robotic arms, and automatic packaging machines, enabling fully unmanned operations. In 2018, the super robot sorting center jointly built by Cainiao and YTO Express was officially put into operation at the YTO Express Hangzhou Transit Center, with 350 robots capable of sorting over 500,000 parcels daily. In 2019, the number of China's intelligent express delivering cabinets reached 406,000, with an annual growth rate of 49.3%, and nearly 26 times that of 2014.

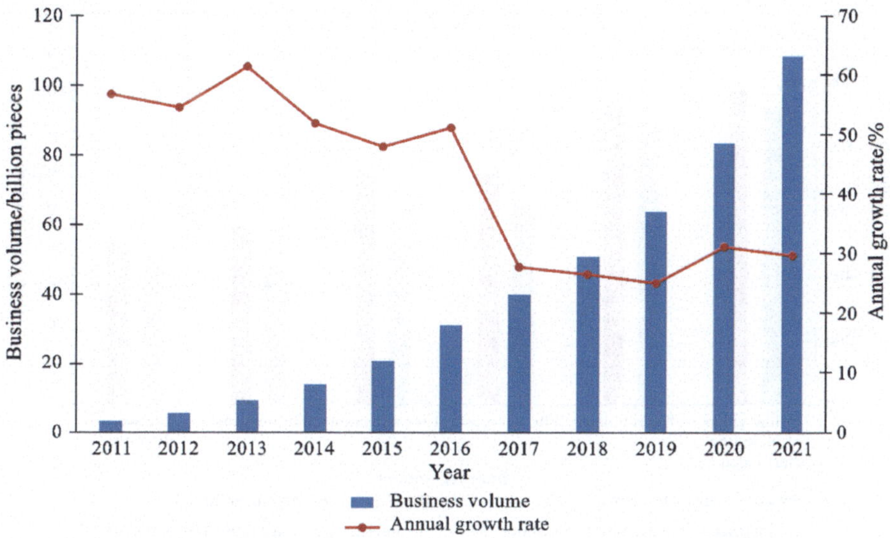

Fig. 4.13 China's express service enterprises business volume and annual growth rate (2011–2021). *Data Source* State Post Bureau of China

4.2.3 Content Services

Content services are the major applications of the Internet. Initially, search engines dominated Internet content services. Recently, to meet the fragmented and multi-scenario reading needs of mobile Internet users, news-pushing applications using "algorithm editing plus intelligent distribution", such as Toutiao, have emerged. As of June 2022, the number of search engine users in China had reached 821 million, a decrease of 7.37 million from December 2021, representing 78.2% of all Internet users; the number of online news users had reached 788 million, an increase of 16.98 million from December 2021, accounting for 75.0% of all Internet users.

The entertainment industry, including music, video, live streaming, and gaming, is also a significant part in Internet content services. As of June 2022, online video users in China had numbered 995 million, an increase of 20.17 million from December 2021, making up 94.6% of all Internet users. Major online video platforms focused on vertical markets by producing original and custom dramas to attract and retain paid members, thus boosting membership revenue. Membership revenue had become a primary revenue source for these platforms, alongside advertising. For instance, in the first quarter of 2022, Bilibili had 20.1 million paid members, with an annual growth rate of 25%, of which nearly 80% were annual or auto-renewal subscribers. iQIYI's membership service revenue reached 4.5 billion yuan, with an annual growth rate of 4%. Figure 4.14 shows the user scale and usage rate of various categories in China's online content (entertainment industry) in December 2020 and December 2021.

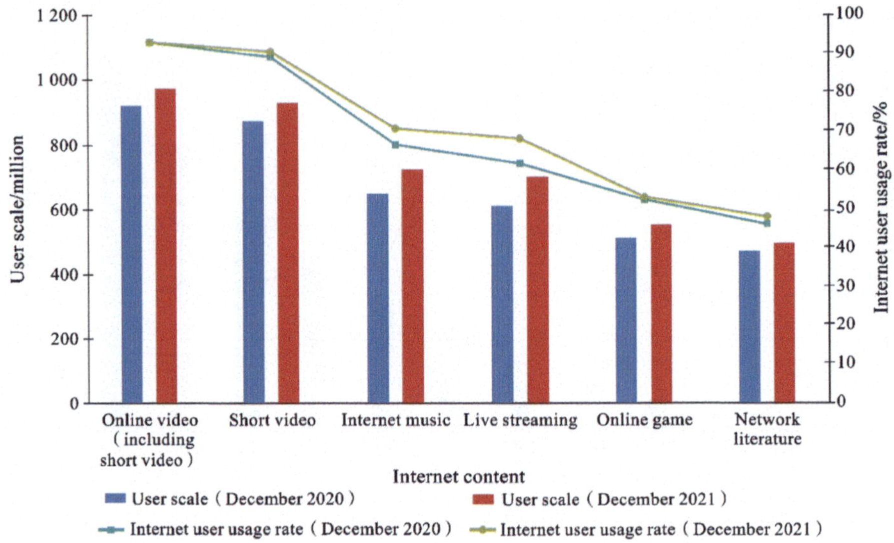

Fig. 4.14 The user scale and usage rate of various categories in China's online content (entertainment industry). *Data Source* China Internet Network Information Center

4.2.3.1 Short Video

The short video industry emerged in 2011, initially centered on personal users creating and sharing homemade videos. This new form of communication suits the fragmented and decentralized nature of mobile Internet, offering diverse content types and enhanced social features. It fulfills users varied content and social needs, becoming a key and rapidly growing sector in the Internet industry.

According to statistics from the CNNIC, by December 2021, the number of short video users in China had reached 934 million, with an annual growth rate of 7%, representing 90.5% of all Internet users. Data from QuestMobile shows that in December 2021, Douyin[2] and Kuaishou had 672 million and 411 million monthly active users, respectively. In December 2021, the average monthly usage time for short videos increased to 53.2 h per person, with total short video usage time accounting for 25.7% of total Internet usage time, surpassing instant messaging as the most time-consuming activity for Internet users. The number of short video users far exceeded that of music, live streaming, gaming, and other entertainment forms, with Internet user penetration rate still rising. Short videos have become a crucial form of online social interaction and entertainment. Figure 4.15 shows the user scale of major short video platforms in China in December 2020 and December 2021.

[2] The TikTok mobile application (released in 2017) is the English international version of the Douyin application (released in 2016).

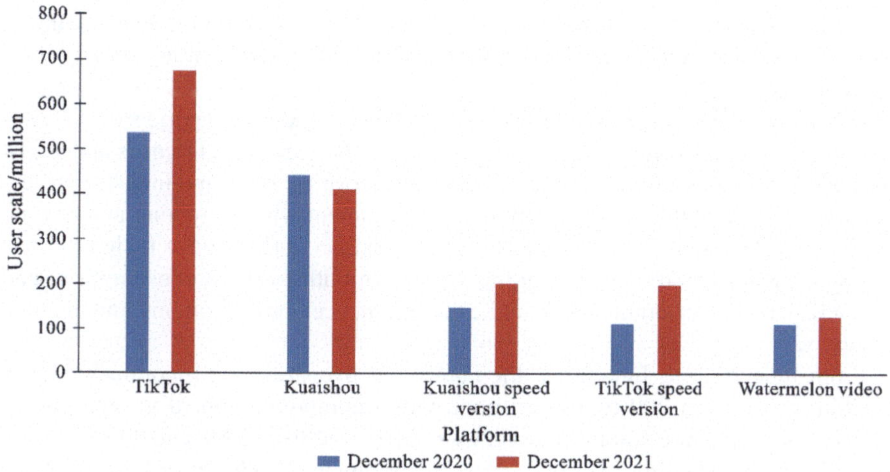

Fig. 4.15 User scale of major short video platforms in China (December 2020 and December 2021). *Data Source* QuestMobile

Short video platforms have a large user base and high active penetration rate, with users spending significant amounts of time on these platforms, accumulating extensive user data. This broadens the audience for advertisements and allows personalized recommendation algorithms to precisely target advertisements, improving conversion rates. Short video advertising leads the online advertising market and has become a crucial channel for digital marketing. In the Chinese Internet advertising market, short video advertising overtook search engine advertising in 2020 with a market share of 17.4%, making it the second largest advertising type after e-commerce advertising.

Short videos have shifted China's e-commerce user behavior from search-based to recommendation-based consumption. Traditionally, online shoppers search for product and service information, but with advancements in big data and artificial intelligence (AI), personalized recommendations based on algorithm have become mainstream. From May 2021 to April 2022, TikTok hosted over 9 million live streams each month, selling more than 10 billion items, with the total transaction value increasing by 2.2 times year-on-year.

4.2.3.2 Live Streaming

The live streaming industry in China began in 2005, initially focusing on vertical and pan-entertainment categories. It has since expanded to include e-commerce, gaming events, pan-entertainment, education, and travel live streaming. In 2016, the women's e-commerce platform MOGU and the comprehensive e-commerce platform Taobao launched live streaming for selling products. Between 2016 and 2018, the major short video platforms Douyin and Kuaishou also quickly entered the live streaming e-commerce space. As of June 2022, China had 716 million online live streaming users, an increase of 12.9 million since December 2021, representing 68.1% of all Internet users.

Live streaming e-commerce offers consumers an immersive experience with real-time interaction, allowing hosts to quickly understand viewer preferences and adjust marketing strategies. A wide range of Internet celebrities, stars, and industry professionals have joined this trend, creating an environment where anything and anyone can be streamed. Online live streaming and social networking have fueled the rise of social media marketing, leveraging word-of-mouth to drive consumption, and giving rise to new economic models like the Internet celebrity economy and the trust economy.

From June 2020 to June 2022, the number of live streaming e-commerce users in China grew from 309 to 469 million, with a compound annual growth rate of 23.2%. The usage rate among Internet users increased from 32.9% to 44.6%, rising by 11.7 percentage points over two years. According to CNNIC, as of June 2022, the number of live streaming e-commerce users in China had reached 469 million, up 5.33 million from December 2021, representing 44.6% of all Internet users. In April 2020, the China Consumers Association released the "Outline Survey Report on Consumer Satisfaction with Live Streaming E-commerce Shopping", which indicated that 68.5% of consumers used Taobao Live, with 46.3% frequently using it. Figure 4.16 shows the market share of consumers on China's live streaming e-commerce shopping platforms from January 2020 to March 2020.

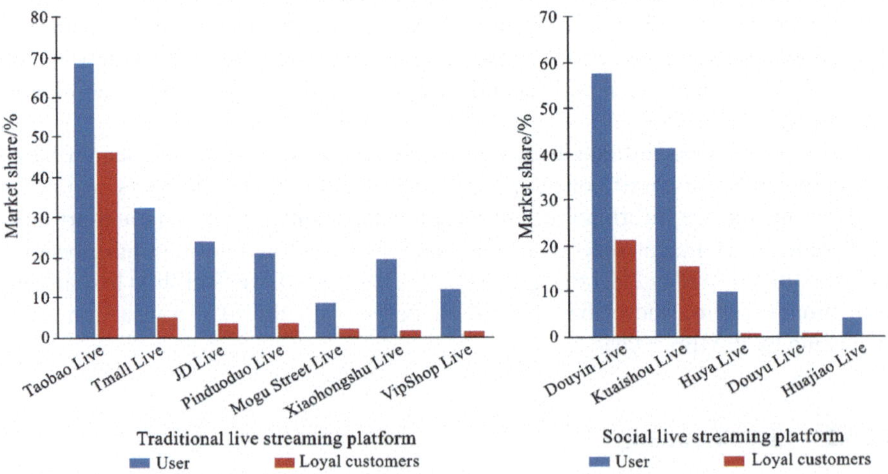

Fig. 4.16 The market share of consumers on China's live streaming e-commerce shopping platforms (January 2020–March 2020). *Data Source* China Consumers Association

4.2.4 *Life Services*

The digitization of the service industry has accelerated evolving in China, with the O2O (online-to-offline) interaction model becoming increasingly vibrant, driving the integration and innovation of the Internet and the service sector. Mobile Internet and other new technologies continuously foster online-offline interaction, offering consumers more service options and delivery convenience. By deeply analyzing user data, enterprises can better match goods and services with user needs, promoting the integration of industry chain resources, the transformation of physical stores, the improvement of operational efficiency, and innovation in business models. O2O is currently one of the most active fields for mass entrepreneurship and innovation. In service sectors such as tourism, car rental and ride-hailing, food delivery, domestic services, beauty and health, education and training, and vehicle maintenance, the variety of products and service models based on the O2O model continue to expand.

Digital technology is rapidly integrating with various industries, driving the recovery of the consumer market by shifting offline advantages in dining, tourism, and shopping to online platforms. As of December 2021, China had 469 million users of online office services, 298 million users of online medical services, and 544 million users of online food delivery, with annual growth rates of 35.7%, 38.7%, and 29.9%, respectively. New business models such as online tourism, adoptive agriculture, and creative homestays are flourishing. By June 2022, the number of online travel booking users in China had decreased to 333 million, a drop of 64.6 million from December 2021, accounting for 31.6% of all Internet users. The number of online medical service users had increased to 300 million, up by 1.96 million from December 2021, making up 28.5% of all Internet users.

4.2.5 *Payment Service*

4.2.5.1 Online Payment

China is a leader in online payment. As of December 2015, the number of online payment users in China had reached 416 million, surpassing online shopping users for the first time, which indicated the Internet's growing penetration into offline consumption scenarios and the emerging habit of going out "wallet-free" and spending "cashless". By June 2022, the number of online payment users in China had reached 904 million, with an increase of 810,000 from December 2021, accounting for 86.0% of all Internet users.

Since 2020, online payment and contactless payment methods have become deeply integrated, making online payment the third largest online application after instant messaging and online video (including short videos) in China. The integration of online and offline consumption has essentially formed. According to the "2020 Mobile Payment User Survey Report" by the Payment & Clearing Association of

China, nearly 75% of users used mobile payment daily. Data showed that in the first quarter of 2022, banks processed 23.57 billion online payment transactions, totaling 585.16 trillion yuan, with annual growth rates of 4.60% and 5.72%, respectively. Mobile payment transactions reached 34.653 billion, totaling 131.58 trillion yuan, with annual growth rates of 6.24% and 1.11%, respectively.

Alipay began its global expansion in 2007. As of June 2019, Alipay and its overseas partners had provided mobile payment and inclusive financial services to over 1.2 billion users worldwide, making it the world's largest non-social application. Alipay offers three main financial services to global merchants and consumers: online payment channels through Global Pay, offline payment for outbound tourism, and global digital inclusive financial services.

Global Pay involves Alipay collaborating with financial institutions and third-party partners worldwide to facilitate online payment channels worldwide, allowing global consumers to complete cross-border payments using localized methods. Nowadays, Alipay partners with over 250 financial institutions globally, facilitating online payment in more than 220 countries and regions and supporting transactions in 27 currencies. Global Pay makes global buying and selling possible. Figure 4.17 shows the user scale of China's online shopping and online payment from June 2007 to June 2022.

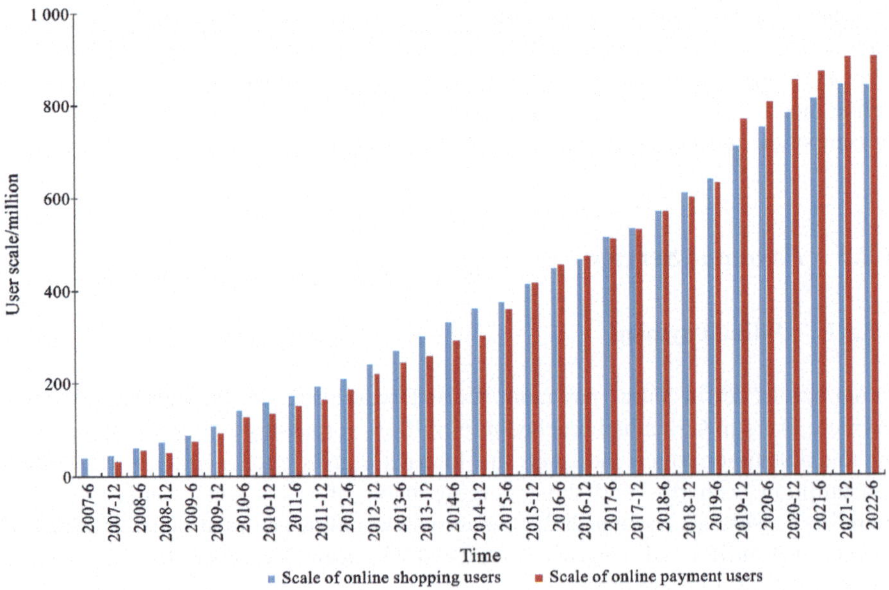

Fig. 4.17 The user scale of China's online shopping and online payment (June 2007–June 2022). *Data Source* China Internet Network Information Center

4.2.5.2 Unmanned Economy

The unmanned economy is a result of advanced ICT, based on the Internet, utilizing big data, Radio Frequency Identification (RFID), IoT, and other sensing technologies to reduce labor output. Unmanned supermarkets, vending machines, and unmanned shelves are the key examples of unmanned economic terminals. China's unmanned retail economy has developed rapidly, with the market scale growing from approximately 1.7 billion yuan in 2014 to about 19.8 billion yuan in 2018, and over 40 billion yuan in 2023. It is expected to surpass 80 billion yuan by 2026, becoming prominent in retail channels. Vending machines remain the dominant channel, holding the largest market share. In 2015, the sales volume of vending machines in China reached 128,200 units, mainly in economic developed areas such as Beijing, Shanghai, Guangzhou, and Shenzhen. The service types have expanded from traditional food vending machines to include self-service coffee machines, beverage machines, and KTV sing-along machines. Figure 4.18 shows China's unmanned retail market scale and annual growth rate from 2016 to 2022.

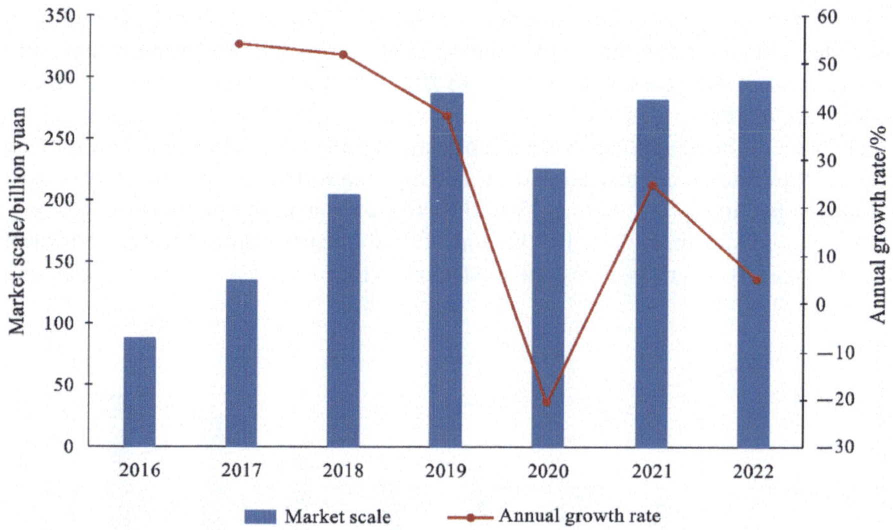

Fig. 4.18 China's unmanned retail market and annual growth rate (2016–2022). *Data Source* Compiled by Insight and Info

4.3 Software and Information Technology Services Industry

China's software and information technology services industry has seen rapid development, with significant growth in scale and benefits. From 2012 to 2021, the scale of the software industry grew approximately 2.8 times, from 2.4794 trillion yuan in 2012 to 9.4994 trillion yuan in 2021, with a compound annual growth rate of 16.1%, ranking among the top industries in the national economy. With its proportion within the ICT industry increasing from 23.4% in 2012 to 33.0% in 2021, the software and information technology services industry has become a key driver for expanding the ICT industry. Figure 4.19 shows the revenue growth of China's software and information technology services industry from 2013 to 2020.

Technology of the People's Republic of China The innovation system for the software and information technology services industry has been established, with significant breakthroughs in basic software such as operating systems, databases, middleware, and office software, achieving a series of landmark achievements. Innovative applications like remote office and collaborative research and development software strongly support various industries. In 2022, China registered 1.835 million computer software copyrights, maintaining over one million registrations annually for five consecutive years. Compared with 2012, the annual number of registrations of software increased 12-fold.

The integration of applications is deepening, significantly enhancing their impact. In 2022, the number of registration of AI and big data software applications exceeded 35,000, respecting a 32-fold and 45-fold increase compared with their initial registration numbers, respectively. By the end of 2020, nearly 100 influential industrial Internet platforms had been established, connecting over 70 million devices and

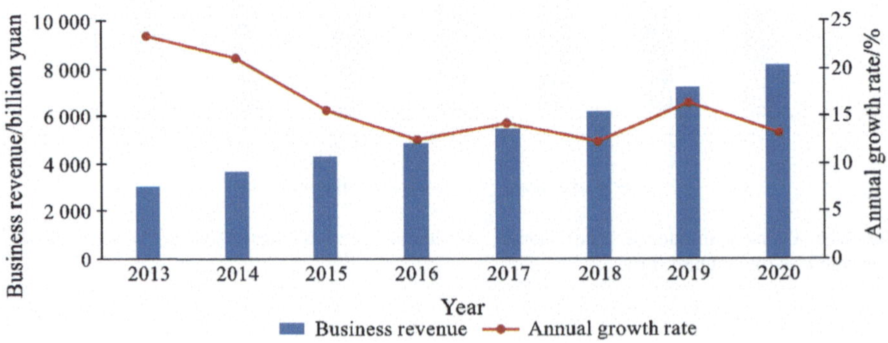

Fig. 4.19 The revenue growth of China's software and information technology service industry (2013–2020). *Data Source* Ministry of Industry and Information

hosting more than 350,000 industrial applications, effectively promoting the transformation and upgrading of the manufacturing industry; over one million enterprises had adopted cloud services, and software and information technology services consumption accounted for more than 50% of the total information consumption.

The strength of key enterprises has grown, and their international competitiveness has been significantly improved. In 2020, China's software and information technology services industry had over 40,000 large-scale enterprises, employing 7.047 million people; the top 100 enterprises generated more than 25% of the industry's total revenue, up five percentage points from 2015, with research and development investment accounting for 27.9% of the industry total; 10 enterprises had revenues exceeding 100 billion yuan, and 2 ranked among the top 10 global enterprises by market value.

The industrial clustering effect has become prominent, and the overall service system for this industry has been more complete. In 2020, 268 software industrial parks contributed over 75% of the total software revenue in China, and 13 renowned software cities accounted for 77.5%. During the 13th Five-Year Plan period, 269 national software standards and 43 industry standards were established, policies benefiting enterprises, such as tax incentives, were further improved, and public service systems for investment and financing, intellectual property, and talent cultivation were continuously optimized.

4.3.1 Fundamental Software

4.3.1.1 Operating System

The operating system is the most critical and fundamental software in a computer, situated above hardware components like the central processing unit (CPU) and memory and beneath various application software, making it the closest software to the hardware layer. Operating systems are classified by the kernel code into open-source systems, such as Linux, and closed-source systems, such as Windows and MacOS. They can also be categorized by the terminal device type into desktop, server, and mobile operating systems. Globally, Linux and Windows dominate the server operating system market. In 2021, of the 13.54 million servers shipped worldwide, 63.47% ran Linux, 36.25% ran Windows, and only 0.28% ran Unix or other systems. In 2022, Linux led China's server operating system market with a market share of 79.1%, followed by Windows with a market share of 20.1%.

The "ZTE Incident"[3] in 2018 underscored the need for China to develop its independent digital ecosystem. As the fundamental layer of computer software, domestic

[3] On the evening of April 16, 2018, the US Department of Commerce announced that the US government would prohibit ZTE from purchasing sensitive products from the US companies for the next seven years. From then on, more and more Chinese high-tech companies, such as Huawei, were prohibited to purchase electrical and related products and parts from the US.

operating systems are vital to creating a secure and independent digital ecosystem for China. Developing these systems is essential. China's domestic operating systems have progressed significantly, moving from being merely "usable" to "user-friendly".

For desktop computers, China has developed over 30 domestic operating systems, most of which are secondary developments based on the Linux kernel. Major manufacturers include Kylin, UnionTech, iSoft, and Euler. These domestic operating systems have been implemented in various sectors, such as party and government, finance, transportation, energy, and telecommunications, and are gradually penetrating core business areas.

For servers, Huawei has developed the Euler open-source operating system (openEuler) tailored for digital infrastructure. OpenEuler is designed for enterprise-level applications such as databases, big data, cloud computing, and AI, supporting Kunpeng and various other processors. As of November 2022, openEuler had been installed on 2.45 million systems, with over 750,000 community users and more than 12,000 contributors in the Euler community.

For mobile terminal devices, Huawei has developed the Harmony operating system (HarmonyOS), which can run on smartphones, computers, tablets, TVs, smartwatches, and other consumer electronic devices and household appliances. According to the globally renowned market research institution Strategy Analytics, HarmonyOS held a 2% market share in global mobile operating systems in 2022, ranking third after iOS and Android. By July 2022, HarmonyOS Connect had over 2000 partners, and more than 320 million Huawei devices were equipped with the HarmonyOS.

4.3.1.2 Database

The database is the collection of related data. A database management system (DBMS) is system software responsible for building, using, and maintaining databases. It provides key functions such as data definition, organization, storage, management, operation, and maintenance through organizing, indexing, querying, and modifying database files. Based on the deployment method, databases can be categorized into traditional private databases (deployed within enterprises or platforms) and cloud databases (deployed on public clouds). According to iResearch, the total market scale of databases in China reached 28.68 billion yuan in 2021, a 16.1% increase from 2020. From 2021 to 2026, the compound annual growth rate is expected to reach 13.4%.

The areas of China's domestic market feature databases in enterprises like Dameng, KingbaseES, GBASE, and Shenzhou General. These databases can also be deployed on the cloud to serve users. With the rise of cloud computing, cloud-native databases have emerged, offering advantages in leveraging cloud environments, such as Alibaba Cloud's PolarDB and Huawei Cloud's GaussDB. According to iResearch, in 2020, databases deployed on public clouds accounted for 32.7% of the market share by revenue in China. In 2021, Alibaba Cloud's distributed database PolarDB entered

Gartner's global database leader quadrant for the first time, ranking third in global cloud database market share.

With the explosive growth of data and increasing application loads, distributed database solutions have become popular for enhancing database access performance. Notable self-developed databases in China include OceanBase and TiDB. OceanBase, a self-developed distributed database initially developed by Alibaba, replaced Oracle in 2014 to support Alipay's core transaction system. OceanBase handled all transaction traffic for the 2015 "Double 11" shopping festival and supported Alipay's core accounting and payment systems in 2016, achieving peaks of 120,000 transactions per second and 175,000 transactions per second. In 2019, OceanBase broke Oracle's nine-year world record in the TPC-C test with 60.88 million tpmC (transations per minute, Concurrent) and set a peak record of 61 million transactions per second during the "Double 11" shopping festival. In 2020, it achieved 707 million tpmC in the TPC-C test.

4.3.2 Cloud Computing

Cloud computing is a kind of pay-by-use model that allows users to access a shared pool of configurable computing resources (including networks, servers, storage, applications, services, etc.) via convenient and on-demand network access. In 2006, Google officially introduced the concept of "cloud computing", with focuses on data services and server infrastructure. Since then, more and more developers and enterprises have adopted cloud computing, leading to the growth of the cloud service industry, primarily centered around IaaS (Infrastructure as a Service), PaaS (Platform as a Service), and SaaS (Software as a Service) delivery models.

The year 2015 marked the beginning of China's cloud computing policy initiatives. China successively issued three policy documents related to cloud computing, covering industry development, industry promotion, application infrastructure, and security management. In April 2022, the MIIT began drafting the "Cloud Adoption Implementation Guide (2022)", aiming to continuously deepen enterprise cloud adoption, further enhance the capabilities and effectiveness of cloud computing applications, and promote high-quality cloud adoption amd application among enterprises.

China's cloud computing market has been growing rapidly. In 2021, the market developed rapidly, with a market scale of 322.9 billion yuan, an increase of 54.4% from 2020. Specifically, the public cloud market scale surged by 70.8% to 218.1 billion yuan, and is expected to drive market growth in the coming years; the private cloud market also surpassed 100 billion yuan, growing by 28.7% to 104.8 billion yuan. China's public cloud IaaS and PaaS segments saw high growth, while SaaS developed steadily. In 2021, the public cloud IaaS market reached 161.47 billion yuan, an 80.4% increase, accounting for nearly 75% of the total market scale. PaaS had the highest annual growth rate at 90.7%, reaching 19.6 billion yuan. The SaaS

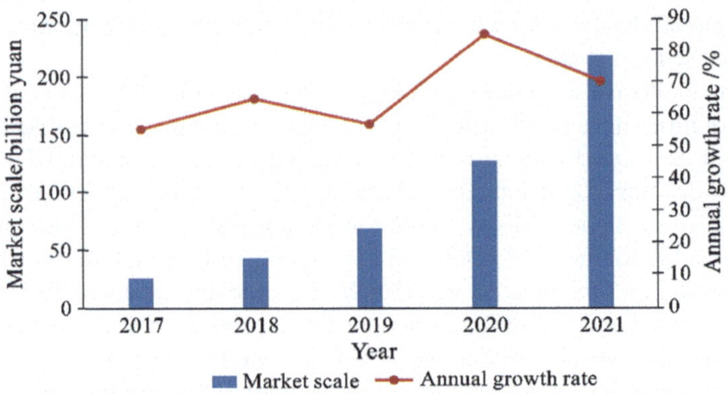

Fig. 4.20 China's public cloud market scale and annual growth rate (2017–2021). *Data Source* China Academy of Information and Communications Technology

market grew steadily to 37.04 billion yuan. Figure 4.20 shows China's public cloud market scale and annual growth rate from 2017 to 2021.

The cloud services provided by Chinese enterprises have witnessed rapid development. By 2021, Alibaba Cloud had deployed hundreds of cloud data centers across 25 global regions, with a total of 80 availability zones. Alibaba Cloud's self-developed cloud operating system, Feitian, manages millions of servers worldwide, covering major overseas markets such as Southeast Asia, Japan, Australia, the US, Europe, and the Middle East. According to the Internet Data Corporation (IDC), Alibaba Cloud ranked third in the global public cloud IaaS market in 2020, following Amazon and Microsoft. Gartner's data indicated that Alibaba Cloud was the leader in the Asia–Pacific region in 2020, with a market share equal to the combined shares of Amazon and Microsoft.

4.3.3 Big Data

Big data refers to the technologies for processing large volumes of data characterized by volume, variety, velocity, and value. The big data industry, which focuses on data generation, collection, storage, processing, analysis, and services, is a strategically emerging industry. In August 2015, the State Council issued the "Action Outline for Promoting the Development of Big Data", elevating big data to a national development strategy. In 2016, the MIIT released the "Development of the Big Data Industry (2016–2020)", leading to a peak in the development of the big data industry. In recent years, the Chinese government has strategically focused on the digital economy, data element market, and the layout of national integrated big data centers, relevant ministries have issued over 20 big data policy documents, local governments have introduced more than 300 related policies, and big data management institutions have been established in 23 provinces, 14 municipalities with independent planning status and sub-provincial cities, creating a collaborative framework for big data development at both the central and local levels.

It should be specifically pointed out that the Chinese government actively explores the integrated development of the information industry and regional economies. Taking Guizhou Province as an example, it is one of the economically underdeveloped provinces in southwest China. This province is located on the Yunnan-Guizhou Plateau, with 92.5% of the area covered by mountains and hills, and the per capita arable land area in Guizhou Province was only 1.35 mu (1 mu≈666.67 square meters). It was a key province for poverty alleviation by the Chinese government. Guizhou's alpine landform provides an environment with cool climate, sufficient electricity, and low seismic activity, which is suitable for the development of computing power infrastructure such as large-scale data centers. To this end, China proposed the development of Guizhou's big poverty alleviation and big data strategy, and promoted the migration of the domestic big data industry to Guizhou.

The National Big Data Comprehensive Experimental Zone has been carrying out experiments and explorations in areas such as big data system innovation, public data open sharing, big data innovative applications, and big data industry aggregation. In September 2015, Guizhou Province launched the construction of China's first National Big Data Comprehensive Experimental Zone. As of 2022, there had been 11 super-large data centers. In 2017, the People's Government of Guizhou Province signed an agreement with Apple, under which the operation of Apple's iCloud service in Chinese mainland would be undertaken by Yunshang Guizhou Company. At the same time, the construction of Huawei, Tencent, and Apple data centers was initiated one after another. In Guizhou Province, Guizhou Gui'an New District became one of the regions with the largest number of super-large data centers in the world. Big data has become an important tool for Guizhou's economic development. In 2021, the revenue of Guizhou's software and information technology services industry saw a year-on-year increase of 59.3%, the fastest growth rate in China. Notably, cloud services became Guizhou's veritable "first industry", accounting for 46.4% of the revenue of software and information technology services industry from 23.1% in 2020. In 2021, the growth rate of Guizhou's digital economy reached 20.6%, 4.4 percentage points higher than the national average growth rate of China's digital economy. As of 2022, Guizhou had gathered more than 9000 big data enterprises with an output value of more than 100 billion yuan, attracting tens of thousands of big data talents every year.

In October 2016, the second batch of big data comprehensive experimental zones was approved, including 2 cross-regional zones (Beijing-Tianjin-Hebei region, Pearl River Delta), 4 regional demonstration zones (Shanghai, Henan, Chongqing, Shenyang), and 1 infrastructure coordinated development zone (Inner Mongolia). Additionally, 11 national new industrialization industry demonstration bases in the field of big data were organized. Local governments and enterprises had also been actively promoting the development of the big data industry, establishing over 100 big data industrial parks.

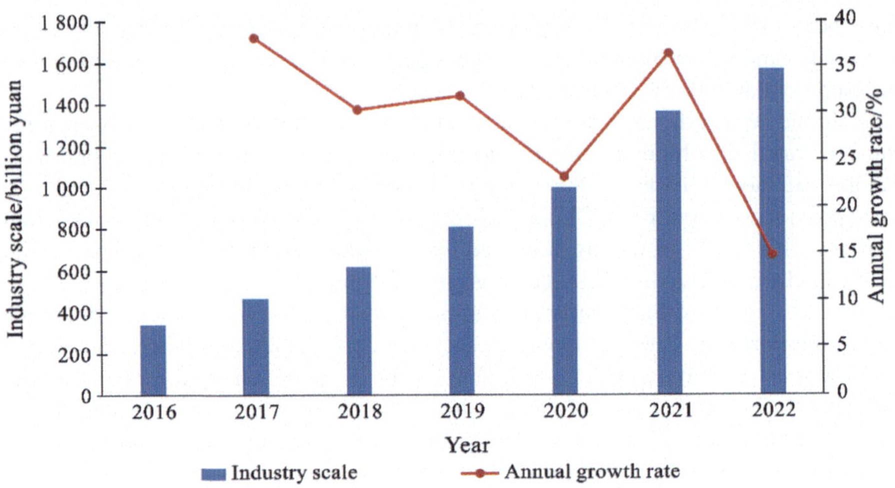

Fig. 4.21 China's big data industry scale and annual growth rate (2016–2022). *Data sources* Cyberspace Administration of China, National Data Bureau of China

From 2017 to 2021, China's data production increased from 2.3 ZB to 6.6 ZB, accounting for 10.5% of the global total and ranking second worldwide in 2021. The big data industry in China grew rapidly, expanding from 340 billion yuan in 2016 to 1.57 trillion yuan in 2022, with a compound annual growth rate of over 30%. The number of provincial public data open platforms increased from 5 in 2017 to 24 in 2021, and the number of open, valid data sets grew from 8398 to nearly 250,000. By 2022, 208 provincial and municipal governments had launched government data open platforms, and the national integrated government data sharing hub had released 15,000 types of data resources, supporting over 500 billion sharing calls. The application scope of big data in China was rapidly expanding, with the industry scale achieving significant growth. Figure 4.21 shows China's big data industry scale and annual growth rate from 2016 to 2022.

4.3.4 Artificial Intelligence

AI is a strategic emerging technology that drives the new wave of technological revolution and industrial transformation. The "Guideline of the State Council on Actively Promoting the Internet Plus Action", issued in July 2015, first listed AI development as a key task. In July 2017, the State Council released the "New-Generation Artificial Intelligence Development Plan", elevating the development of AI to a national strategy. In 2019, the seventh meeting of the Central Committee for Deepening Overall Reform approved the "Guideline on Promoting the Deep Integration of Artificial Intelligence and the Real Economy", aiming to leverage AI's integrative nature to advance industrial upgrading. The MIIT issued the "Three-Year Action Plan for Promoting the Development of a New Generation of Artificial Intelligence Industry (2018–2020)" and other policy documents. Additionally, more than 20 provinces, autonomous regions, and municipalities have issued over 60 special plans for AI.

China's innovation capability on AI technology continues to improve. China has a relatively high entry rate in global competitions for intelligent tasks such as computer vision and speech, and has repeatedly refreshed the accuracy record for SOTA model (state-of-the-art model, the best model on one task) of intelligent tasks in global competitions, such as conversational question and answer, reading comprehension, and face recognition. The total number of AI papers, the number of highly cited papers, and the number of invention patents from China are in the first echelon. The AI patents from China increased from 7968 in 2012 to 80,785 in 2021, accounting for 70.9% of the world's total, ranking first in the world. The published papers on AI technologies in China had increased from 3423 in 2012 to 26,000 in 2021, accounting for 26.5% of the world's total and ranking first in the world. From the perspective of the subdivision of papers, the themes of China's AI papers are mainly concentrated in the fields of decision-making systems, computer vision, deep learning, intelligent robots, expert systems, fault diagnosis, and neural networks.

The scale of the AI industry in China has been continuously expanding. The industrial ecology has been basically formed, and the overall strength of the industry has been significantly enhanced. As of 2020, the scale of China's AI industry, the number of key enterprises (including the number of unicorn enterprises) were both second only to the US, ranking second in the world, and a relatively complete industrial chain had been established.

In terms of AI computing power infrastructure, China has built the world's largest AI computing power network cluster. Several AI computing centers have been established and put into operation: Pengcheng Cloud Brain II, Hengqin Advanced Intelligent Computing Center, Wuhan AI Computing Center, Nanjing AI Computing Center, Nanjing Kunpeng/Shengteng AI Computing Center, Xi'an Future AI Computing Center, Central Plains AI Computing Center, and Chengdu AI Computing Center, etc. Among these centers, the Hengqin Advanced Intelligent Computing Center, built in Zhuhai, Guangdong Province in 2019, is the world's largest AI computing center. The "Intelligent Computing Network", jointly developed by Pengcheng Laboratory and Huawei in June 2022, is a core segment of the first phase of the China's first AI computing power network, marking the full launch of the "China Computing NET" plan. The world's largest integrated AI computing power network has gradually taken shape.

In terms of core AI technologies, several Internet enterprises in China have conducted research on a variety of AI technologies. They have established deep learning institutes, AI labs, and intelligent driving teams, focusing on machine learning, deep learning, image recognition, speech recognition, autonomous driving, etc. Over the past decade, Baidu has launched AI products such as "Intelligent Cloud" and "Baidu Brain". Alibaba began building its AI research institute in 2012 and launched the visual AI platform DT-PAI that integrates its core AI algorithms in 2015. Alibaba also introduced the virtual assistant "Ali Xiaomi" and the ET robot.

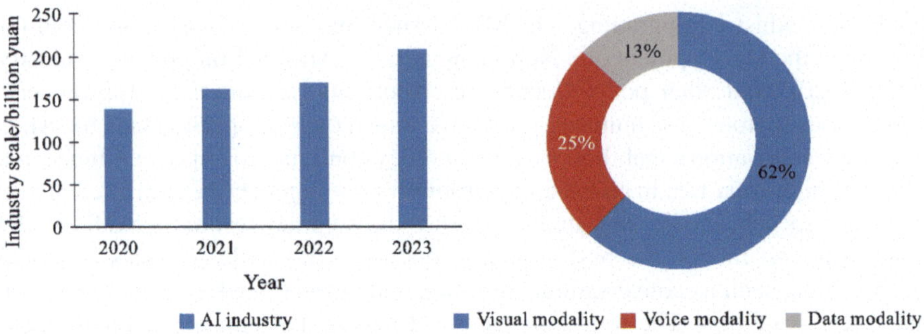

Fig. 4.22 The scale of China's AI industry (2020–2023) and the share of different AI modalities in 2023. *Data Source* iResearch

Additionally, Alibaba has integrated AI with its existing e-commerce, big data, cloud computing businesses and other original businesses, creating intelligent products like warehouse robots, City Brain, E-commerce Brain, Industrial Brain, Alibaba Green Net, and Medical Brain.

In terms of industrial innovation of AI, as of 2022, China had established 8 National New-Generation Artificial Intelligence Innovation Application Pilot Zones and 18 National New-Generation Artificial Intelligence Innovation Development Experimental Zones. The development of AI industry in China has been continuously deepening, showing initial success in integrating with the primary, secondary, and tertiary industries. AI has been expanded from the leading sectors (like healthcare, transportation, and education) to the traditional sectors of manufacturing and agriculture. Intelligent finance, healthcare, security, and transportation have become key application areas for accelerating the AI industrialization. New intelligent products, business models, and format are continuously emerging. Figure 4.22 shows the scale of China's AI industry from 2020 to 2023 and the share of different AI modalities in 2023.

4.3.5 Blockchain

Blockchain is a distributed digital ledger that records information and data, stored across multiple participants in a peer-to-peer network. Participants use cryptographic signatures to add new transactions to the existing chain, creating a secure, continuous, and immutable data structure. Blockchain ensures multi-node consensus, transparency, and tamper-proofing, making it an essential technology for protecting data assets. Developing blockchain technology can promote data element rights confirmation, innovate data sharing and openness methods, and establish a robust data regulatory governance system.

Since China listed the development of blockchain in the "Outline of the 13th Five-Year Plan for National Economic and Social Development of the People's Republic of China" in 2016, the country has continuously introduced policies to support the blockchain industry and encourage the application of blockchain technology across various sectors. In the "Outline of the 14th Five-Year Plan for National Economic and Social Development and the Long-Range Objectives Through the Year 2035 of the People's Republic of China", blockchain was listed as one of the seven key industries in the digital economy. The blockchain related research in China has been in the leading position in the world. According to the list of Top 50 Blockchain Papers and Scholars of 2018 presented by the Silicon Valley Insight Research, 30% of the scholars on the list were from the US, and 28% were from China. In 2018, the global blockchain patent applications reached 2966, with China accounting for 2435, representing 82.1% of the global total.

The scale of the blockchain industry in China has been steadily growing. Since 2020, provincial governments have been actively promoting blockchain technology, initiating over 90 blockchain projects primarily in government services, justice, traceability, and trade finance, with the application scale continuing to expand. In 2021, the industry scale continued to rise, reaching an annual industry size of approximately 6.5 billion yuan, with an annual growth rate of 34.9%. For example, Hainan Province's blockchain-based "financial electronic invoice" application platform ensures that invoices are trustworthy, verifiable, secure, and supervisable. These invoices enhance tracking and auditing of financial expenditures in public services, preventing duplicate reimbursements. By October 2021, Hainan Province had issued 26.61 million blockchain financial electronic invoices, amounting to over 40 billion yuan, with 2795 units online, covering 14 fields such as justice, real estate, education, and healthcare. Figure 4.23 shows China's blockchain industry scale and annual growth rate from 2018 to 2021.

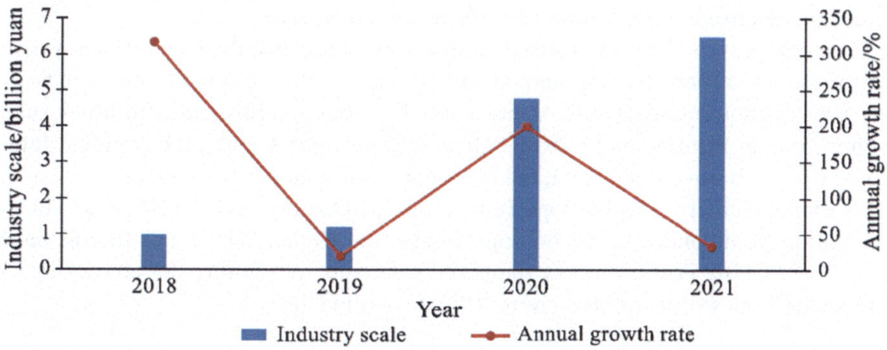

Fig. 4.23 China's blockchain industry scale and annual growth rate (2018–2021). *Data Source* China Center for Information Industry Development

4.4 Electronic Information Manufacturing Industry

From 2012 to 2021, the profits of China's electronic information manufacturing enterprises saw remarkable growth, with the annual growth rate increasing from 7.9% in 2012 to 38.9% in 2021; the operating revenue of China's large-scale electronic information manufacturing enterprises increased from 10.7 trillion yuan to 14.1 trillion yuan, surpassing the average growth rate of China's industrial sector. The contribution of this industry to China's industrial production has been significantly strengthened.

From the perspective of electronic product research and development (R&D), China's electronic information manufacturing industry has made continuous breakthroughs in key components and core technologies, with some areas reaching advanced levels in the industry. In the chip sector, the Loongson 3A5000's performance has approaching that of mainstream desktop CPUs. In the storage sector, a production line for the first-generation 10-nm 8 GB DDR4 product has been established, keeping pace with international DRAM products. In the printer sector, significant progress has been made in laser printing engine core technology, and printing equipment based on Loongson and Phytium chips has been mass-produced and widely used.

From the perspective of technical innovation, the number of invention patent applications of China in the electronic information manufacturing industry has increased rapidly, far exceeding other manufacturing industries. In 2020, it accounted for 27.4% of the number of invention patent applications in the country, which was 14 percentage points higher than that of the second-ranked electrical machinery and equipment manufacturing industry. In the 2021 TOP 10 list of global international patent application enterprises released by the World Intellectual Property Organization (WIPO), Chinese enterprises Huawei, OPPO and BOE were on the list, all of which are electronic information manufacturing enterprises.

From the perspective of industrial cooperation platforms, the Chinese innovative platforms such as new displays and integrated circuits play an important role in tackling key technology problems. At present, China has established institutions such as the National Printing and Flexible Display Innovation Center, the National Integrated Circuit Innovation Center, the National Intelligent Sensor Innovation Center, the National Center of Technology Innovation for Display, and the National Third-Generation Semiconductor Technology Innovation Center. These institutions have rallied the "China Power" in key industrial segments, promoting the innovation of core technologies of integrated circuit (IC) design in China.

4.4.1 Semiconductor Sector

Since 2015, China has intensified its support for the semiconductor industry, with policies and funding being gradually implemented. The "National Outline of Promoting the Development of Integrated Circuit Industry" set policy goals for

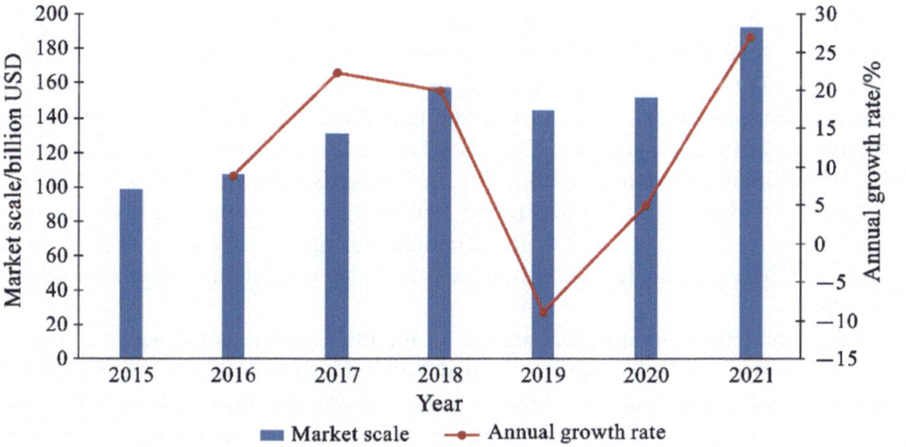

Fig. 4.24 China's semiconductor industry market scale and annual growth rate (2015–2021). *Data Source* Insight and Info

various segments of the semiconductor industry, including the market scale, materials and equipment, design, manufacturing, and packaging, and offered policy support and tax incentives. The 14th Five-Year Plan emphasized that high-end chips are a "national urgent need and long-term demand", and that integrated circuits are a "fundamental core field crucial to overall development and national security".

According to the NBS, China's semiconductor IC production reached 359.4 billion units in 2021, a 33.3% increase from the previous year. According to the data of the Semiconductor Industry Association of America, Chinese enterprises market share in the global semiconductor market was 15% in 2021. In 2021, the market scale of China's semiconductor industry was 192.5 billion USD (1.39 trillion yuan), with an annual growth rate of 27.1%. Figure 4.24 shows the market scale and annual growth rate of China's semiconductor industry from 2015 to 2021.

4.4.1.1 Semiconductor Material

Semiconductor materials refer to special electronic functional materials, interconnection and packaging materials, processes and auxiliary materials used in the preparation of electronic components, modules and systems. These materials are essential throughout the semiconductor manufacturing process. In wafer manufacturing, significant quantities of silicon wafers, electronic specialty gases, photomasks, and polishing materials are used, while packaging and testing primarily involve packaging substrates and lead frames. Semiconductor materials are crucial in the production of electronic information products, with their process level and quality directly determining the performance of components.

In 2020, the NDRC and other departments issued the "Guiding Opinions on Expanding Investment in Strategic Emerging Industries, Fostering New Growth Areas", aiming to accelerate breakthroughs in high-strength and high-conductivity heat-resistant materials, corrosion-resistant materials, large-size silicon wafers, and electronic packaging materials. China's share of the global semiconductor material market increased annually from 2012 to 2019, rising from 12.28% to 16.67%, ranking third globally. According to the National Bureau of Statistics, the revenue of China's electronic specialty material manufacturing industry grew steadily from 2018 to 2020, reaching 160.94 billion yuan in 2020, with an annual growth rate of 39.39%.

Semiconductor silicon wafers are one of the main materials for wafer manufacturing, and the finished product is called "silicon wafers", which are crucial for manufacturing semiconductors such as integrated circuits. From 2016 to 2018, the market sale of Chinese mainland's semiconductor silicon wafer industry grew from 500 million USD to 996 million USD. China is the largest producer of small and medium-sized silicon single crystals and has led the world in the production of printed circuit boards, copper-clad laminates, magnetic materials, and organic films for three consecutive years. From 2017 to 2020, China invested in silicon wafer production projects and expanded silicon wafer production capacity, with wafer plant capacity growing at a compound annual growth rate of 13%. China's wafer production increased from 2.3 million wafers per month in 2015 to 4.717 million per month in 2020, with an expected 8.583 million wafers per month by 2035. According to Knometa Research, in 2021, Chinese mainland's wafer capacity accounted for 16% of the global total, with the expansion speed of wafer plants in Chinese mainland significantly higher than that of other countries and regions. Figure 4.25 shows China's semiconductor material market scale and annual growth rate from 2015 to 2021.

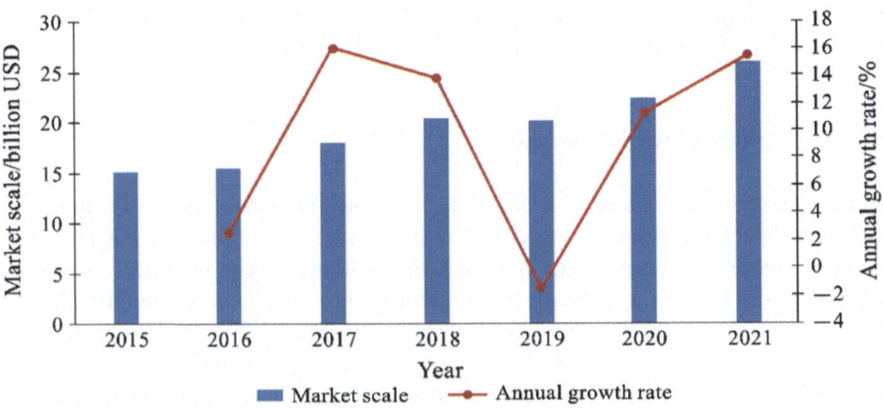

Fig. 4.25 China's semiconductor material market scale and annual growth rate (2015–2021). *Data Sources* International Semiconductor Industry Association, qianzhan.com

4.4.1.2 Semiconductor Equipment

Semiconductor equipment, used in the manufacturing and testing processes of chips, encompasses various types of devices, including lithography machines, etching machines, thin-film deposition equipment, ion implantation machines, testing machines, sorting machines, probe stations, and more. Semiconductor equipment needs to stay ahead of semiconductor product manufacturing development. With each new generation of product, there is a need for a more advanced generation of process equipment. Currently, the global semiconductor equipment market has a high level of concentration, with the top 10 enterprises, primarily from the US, the Netherlands, and Japan, monopolizing over 90% of the global market share.

China places significant emphasis on developing semiconductor equipment. The Ministry of Science and Technology (MOST) launched a major project, "Manufacturing Equipment and Complete Process for Very Large Scale Integrated Circuits", bringing together over 200 entities, including upstream and downstream enterprises, research institutions, and universities involved in manufacturing process, equipment, related components, and materials, to collaboratively advance industries, universities, and research institutions. Domestically produced equipment now covers all stages of semiconductor manufacturing, excelling particularly in etching, cleaning, and thin film equipment. Chinese enterprises NAURA Technology Group Co., Ltd ranks among the Top 10 global semiconductor equipment enterprises. According to the International Semiconductor Industry Association, Chinese mainland's semiconductor equipment market scale reached 29.6 billion USD in 2021, representing 29% of the global market, making it the largest market worldwide. The localization rate of semiconductor equipment in China increased from 17% in 2020 to 20% in 2021. The China Electronic Production Equipment Industry Association reported that in 2021, major domestic semiconductor equipment market scale in Chinese mainland reached 38.6 billion yuan, with a localization rate of 20.2%. Figure 4.26 shows China's semiconductor equipment market scale and annual growth rate from 2016 to 2022.

4.4.1.3 Electronic Components

Electronic components are essential parts of electronic devices and small machines and instruments, and are the general term for electronic devices such as capacitors, transistors, hairsprings, and springs. Electronic components are the cornerstone that supports the development of the information technology industry, and are crucial for ensuring the safety and stability of the industrial chain and supply chain. Taking multilayer ceramic capacitor (MLCC) as an example, the average number of MLCC used per smartphone exceeds 1000, and each new energy vehicle uses more than 10,000.

China has developed the world's largest and most comprehensive electronic component industrial system, with a nearly complete industrial chain. It leads globally in the production of various electronic components, including electroacoustic

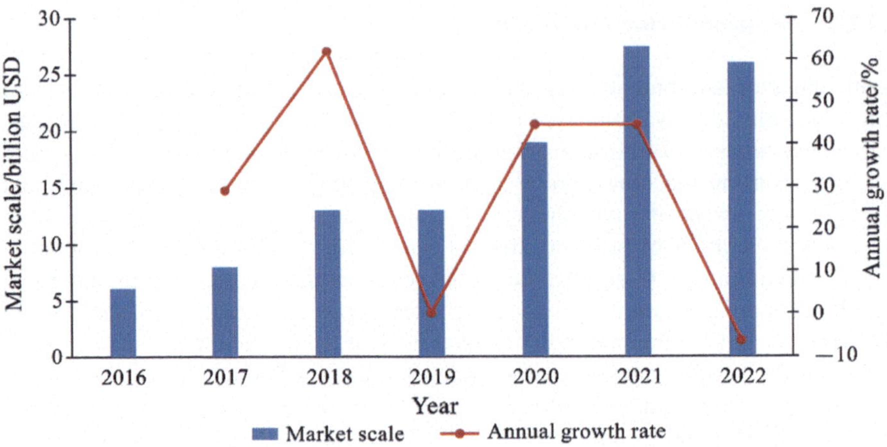

Fig. 4.26 China's semiconductor equipment market scale and annual growth rate (2016–2022). *Data Source* huaon.com

devices, magnetic material components, and optoelectronic cables. The industry's overall scale has exceeded two trillion yuan. In December 2019, the annual growth rate of the added value of China's electronic device and electronic specialty material manufacturing industry was 20.7%, while the annual export delivery value decreased by 2.3%. Among the main products, the annual growth rate of the output of electronic devices was 26.9%. In 2019, the annual growth rate of the operating revenue of this industry was 0.3%, while the annual profit declined by 2.1%. Figure 4.27 shows the output scale and annual growth rate of semiconductor discrete devices in China from 2015 to 2020.

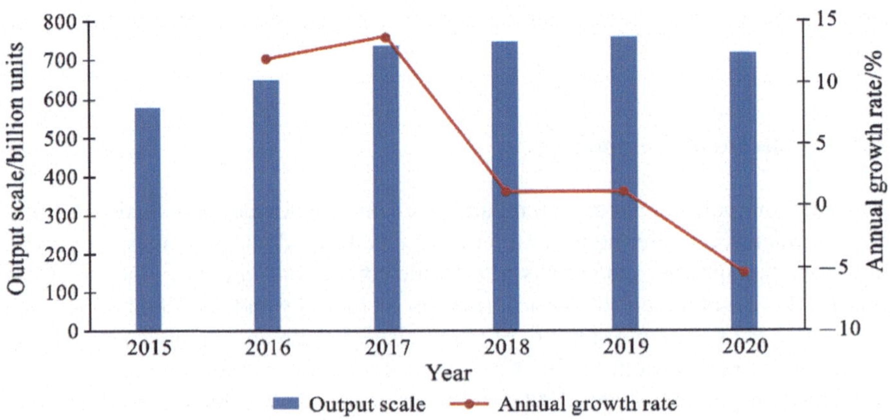

Fig. 4.27 China's semiconductor discrete devices output scale and annual growth rate (2015–2020). *Data Sources* National Bureau of Statistics of the People's Republic of China, qianzhan.com

4.4.1.4 Integrated Circuit and Chip

Semiconductors are typically classified into four categories: ICs, discrete devices, optoelectronic devices, and sensors. ICs are miniature electronic circuits integrated onto a small semiconductor chip through specific design techniques and semiconductor processing technology. IC accounts for about 80% of the global semiconductor sales, as the "industrial food", they are the core of the modern information technology industry. The IC industry mainly comprises design, manufacturing, and packaging and testing sectors.

In 2014, the State Council issued the "National Outline of Promoting the Development of Integrated Circuit Industry", making the development of the IC industry a national strategy. In September of the same year, the National Integrated Circuit Industry Investment Fund was established with an initial capital of over 130 billion yuan. In 2020, the State Council released the "Policies for Promoting the High-quality Development of the Integrated Circuit Industry and Software Industry in the New Era".

Although China's IC industry started relatively late, it has become a key player in the global IC market. According to the China Semiconductor Industry Association, as of the first half of 2020, sales in China's IC industry reached 353.9 billion yuan, with an annual growth rate of 16.1%. From 2018 to 2021, the compound annual growth rate was 17%, over three times the global growth rate during the same period. In 2020, China's IC output reached 261.26 billion units, with an annual growth rate of 29.45%. Figure 4.28 shows China's IC output scale and annual growth rate from 2015 to 2021.

China has made significant strides in enhancing its independent capabilities. With the implementation of national science and technology major projects, the localization verification efficiency of key IC equipment for mainstream process nodes

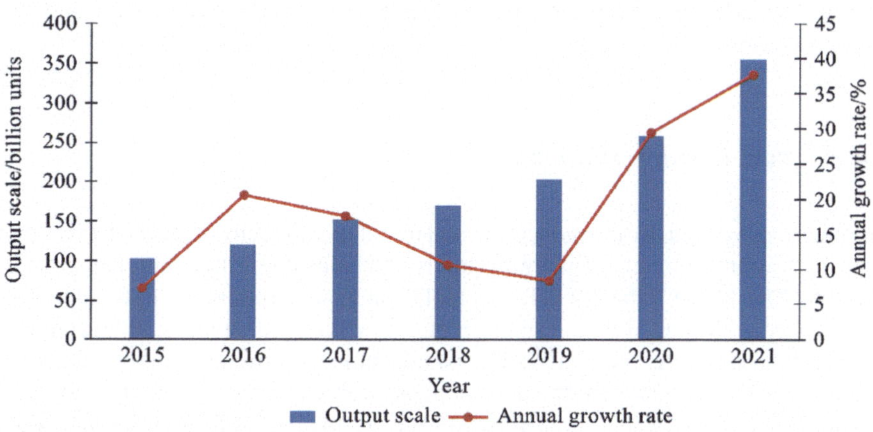

Fig. 4.28 China's IC output scale and annual growth rate (2015–2021). *Data Source* National Bureau of Statistics of the People's Republic of China, qianzhan.com

has increased fourfold. Major equipment such as chemical mechanical polishing machines, dielectric etching machines, and cleaning machines has entered the verification stage for advanced process nodes. The proportion of domestically sourced advanced packaging equipment has reached 79%, saving over 30% in procurement costs. The variety coverage of hundreds of key process materials across seven categories has exceeded 25%, with a localization rate of over 20%.

China's chip design capabilities and innovation levels have also been steadily advancing. As of December 2021, the number of chip design enterprises in Chinese mainland had grown from 2218 in 2020 to 2810, with an annual growth rate of 26.7%. In 2020, high-performance processors were widely used in the party, government, and military markets, and China achieved significant breakthroughs in developing large-capacity, high-density 3D NAND flash memory and fourth-generation DDR4 memory products.

In the sector of high-end chips, China has made substantial advancements. Over the past decade, China's domestically produced general-purpose CPUs have evolved from "barely usable" to "fully functional". Domestic embedded CPUs competed directly with foreign products, achieving annual sales of several hundred million units. Domestic 3D flash memory and DRAM have entered mass production, approaching international advanced technology levels. Additionally, domestic FPGA chips have been fully integrated into the communications and entire machine markets. In the field of domestic electronic design automation tools, significant progress have been made, with the development of a series of key tools for both analog and digital circuit design processes.

In the sector of mid-and low-end chips, China have successfully achieved import substitution, fully replacing European and American products. For instance, Jiangsu Runshi, after four to five years of development, has made significant strides in analog chips. Its products, such as operational amplifiers, comparators, analog switches, and level shifters, can replace those from enterprises like Texas Instruments Inc.(TI), Analog Devices, Inc.(ADI), Microchip Technology Inc.(MICROCHIP), and ON Semiconductor.

4.4.2 New Display Industry

The new display industry is a pioneering and foundational sector. "The 12th Five-Year Plan for National Strategic Emerging Industries" elevated it to a national strategic industry. The NDRC, the MOST, the MIIT, and other ministries have supported the industry with special plans, highlighting key development areas and technological breakthroughs. As the world's largest consumer electronics market, China has also provided momentum for the development of this industry. From 2012 to 2021, China's new display industry transformed from "following" and "running alongside" to "leading", solving the long-standing "few screens" issue. The industry scale and TFT-LCD capacity have both reached global leadership. Figure 4.29 shows China's new display industry scale and annual growth rate from 2017 to 2022.

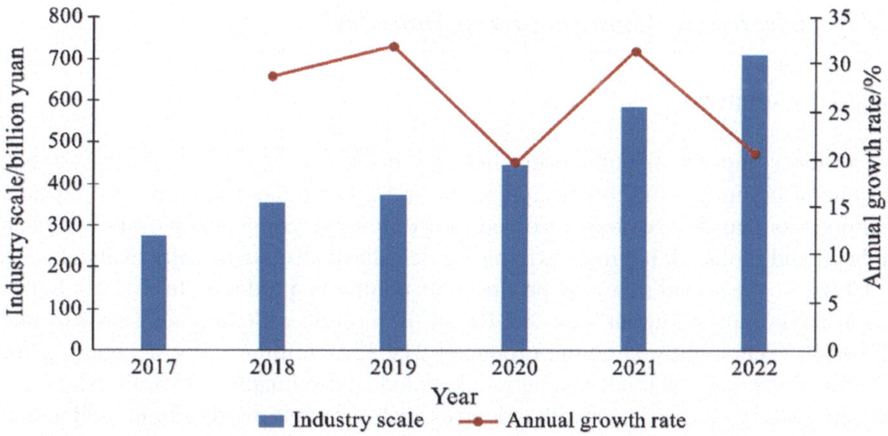

Fig. 4.29 China's new display industry scale and annual growth rate (2017–2022). *Data Source* China Optics and Optoelectronics Manufactures Association

China has developed a complete new display industry chain centered on panel manufacturing. The industry's total revenue grew from 74 billion yuan in 2012 to approximately 510 billion yuan in 2021, with a compound annual growth rate of over 20% over the past decade. In 2021, the overall revenue from China's display support field surpassed 100 billion yuan, ensuring a secure and stable supply chain for the new display industry. China leads the global market in five major panel applications: mobile phones, televisions, tablets, laptops, and monitors. From 2013 to 2022, China's average annual shipments of smartphones, tablets, and television reached 391 million units, 22.03 million units, and 45.95 million units, respectively.

Significant breakthroughs have been made across various technical routes in the new display industry. In terms of intelligent property rights, China has become the largest source of panel technologies, and as of 2021, the number of applications of China's panel patents had accounted for 35% of the global total. Enterprises like BOE, Wuhan CSOT, and Shenzhen CSOT were in the top 50 global international patent applicants in 2021. Technological innovations such as mini-LED and Dual Cell backlight adjustment, and quantum dot color solutions have achieved large-scale production in large-size LCD products. Technologies like low-power adaptive frequency modulation, under-screen fingerprint recognition, under-screen cameras, and Pol-less display have been successfully integrated into medium-and small-size OLED products. Domestically produced flexible OLED screens have entered the supply chains of several global high-end flagship brands.

4.4.3 Electronic Manufacturing Industry

4.4.3.1 Computer

Electronic computer manufacturing industry produces various computer systems, peripheral equipment, terminal equipment and other related devices. As physical components and devices have evolved, not only have computer mainframes been updated and replaced, but their external devices have also continually evolved.

China is the world's largest producer of computer products. In 2021, China's electronic computer output was 485.464 million units, with an annual growth rate of 19.8%. The southwest China produced 219.2221 million units, accounting for 45.05% of the national total. The output included approximately 230 million laptops, 70 million desktops, and many other devices such as tablets, workstations, and industrial control terminals. According to the International Data Corporation, the top 5 vendors in China's electronic computer market in 2021 were Lenovo (32.8%), HP (15.6%), Dell (14.0%), Huawei (10.4%), and Apple (7.3%). Figure 4.30 shows China's electronic computers output scale and annual growth rate from 2016 to 2021.

In the server market, China's server shipments have been steadily increasing, reaching an estimated 3.91 million units in 2021, which was 28.9% of the global total, with an annual growth rate of 11.7%. Inspur, Lenovo, and Huawei are the leading server manufacturers in China. Inspur has become a core member and standard advocate of the three major global open computings organizations: ODCC, OCP, and OPEN19 Fundation. With a business presence in 120 countries and regions, Inspur held a 9.4% share of the global server market in 2021, ranking second worldwide.

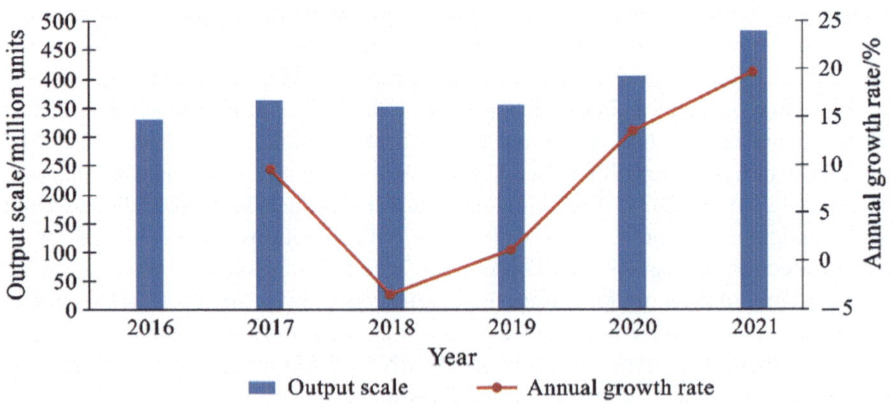

Fig. 4.30 China's electronic computers output scale and annual growth rate (2016–2021). *Data Source* National Bureau of Statistics of the People's Republic of China

4.4.3.2 Communications Equipment

The communications system equipment manufacturing industry involves producing equipment necessary for constructing communications systems, including fixed and mobile communications access, transmission, and switching equipment. This includes providing transmission, access, and bearer network solutions for operators and enterprise customers. Communications terminal equipment includes audio, graphic image, video, data, and multimedia communications terminals.

In terms of communications system equipment, several government departments have issued policies to support and regulate the development of the communications equipment manufacturing industry in recent years. These policies address 5G network construction, terminal IPv6 upgrades, "double gigabit" network infrastructure, and industrial Internet construction. According to the "China Electronic Information Industry Statistical Yearbook 2019", from 2014 to 2019, the operating revenue of China's large-scale communications equipment manufacturing industry continuously grew. The MIIT reported that in 2020, the operating revenue of China's communications equipment manufacturing industry grew by 4.7% year-on-year.

In terms of communications terminals, China has the world's most complete mobile phone industry chain. Over the decade from 2012 to 2021, Chinese enterprises focused on terminal equipment manufacturing, leading to the rapid rise of domestic mobile phone brands. In 2012, China produced 1.18 billion mobile phones, primarily using a low-price strategy. By 2021, this number grew to 1.66 billion, making China the world's largest producer of information and communications terminals. According to Counterpoint Research, China accounted for 67% of the global mobile phone production in 2021. Xiaomi ranked third globally with shipments of 1.903 million units, capturing a global market share of over 14% and capturing three of the top five positions in the global smartphone market along with OPPO and Vivo. According to the International Data Corporation, China's smartphone shipments reached 340 million units in 2021, with an annual growth rate of 13.33%. The China Academy of Information and Communications Technology reported that domestic brand shipments totaled 304 million units, with an annual growth rate of 12.6%, and 5G phones accounted for 266 million units, with an annual growth rate of 63.5% in the same period, making up 75.9% of total shipments. Figure 4.31 shows China's smartphone shipments and annual growth rate from 2011 to 2021.

Huawei (including Honor) used to hold a market share of 46%; however, affected by the trade war initiated by the US in 2018, Huawei's smartphone market share was significantly impacted. The brand of Honor was then forced to be independent. In response to the US chip supply cut-off, Chinese smartphone manufacturers began to strengthen technological innovation and started developing their own chips. Vivo introduced the V1 imaging chip, OPPO launched a 6 nm NPU chip for imaging, and Xiaomi released the Surge C1 and Surge P1 to enhance imaging and charging performance. According to International Data Corporation, the top five smartphone vendors in China by shipments in 2021 were Vivo, OPPO, Xiaomi, Apple, and Honor.

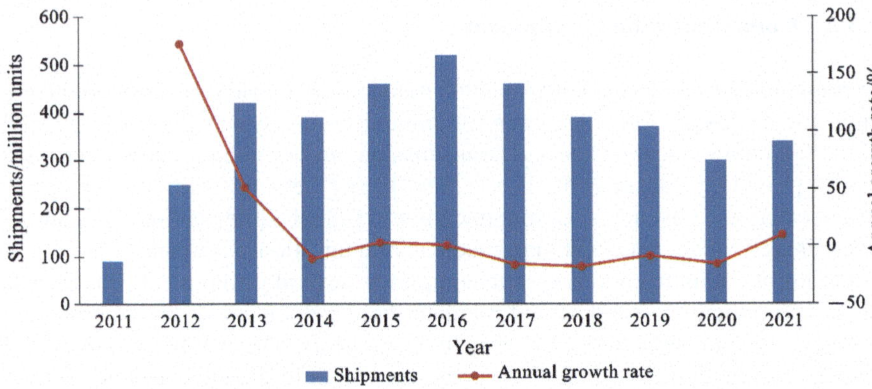

Fig. 4.31 China's smartphone shipments and annual growth rate (2011–2021). *Data Sources* Ministry of Industry and Information Technology of the People's Republic of China, International Data Corporation

4.4.3.3 Consumer Electronics

From 2017 to 2022, China held the position of the world's largest goods trading nation for six consecutive years, leading the world in the production of over a hundred types of products. In 2021, China's home appliance industry generated prime operating revenue of 1.73 trillion yuan, with an annual growth rate of 15.5%, and profits of 121.8 billion yuan, with an annual growth rate of 4.5%. Sub-sectors such as refrigeration products, air conditioners, and kitchen appliances all saw double-digit growth in prime operating revenue. In 2021, China's home appliance export value exceeded 100 billion USD, with products sold in over 160 countries and regions, reaching more than 2 billion households worldwide. The global market share of Chinese-made refrigerators, air conditioners, washing machines, televisions, and small home appliances popular among young people all exceed 50%.

The home appliance industry's development trend was shifting from single-product intelligence to whole-house intelligence. From 2012 to 2021, the market scale of China's smart small appliances grew from 262.1 billion yuan to 379.3 billion yuan, with an annual growth rate of 45%. Wearable device shipments soared from 2.3 million units to 140 million units, a 60-fold increase. In 2020, China's smart home market shipped about 201 million units, and in the first half of 2021, smart home device shipments reached approximately 100 million units, with an annual growth rate of 13.7%. The International Data Corporation predicted that China's smart home device market would continue to grow, with shipments reaching 540 million units by 2025. Figure 4.32 shows China's home appliance retail sales and annual growth rate from 2012 to 2021.

China has the world's largest home appliance manufacturing base. In 2020, Haier operated 14 R&D centers, 122 manufacturing centers, and 108 marketing centers globally, forming an extensive R&D resource-sharing network that includes shared modules, reusable technologies, and some shared patents. Haier Smart Home topped

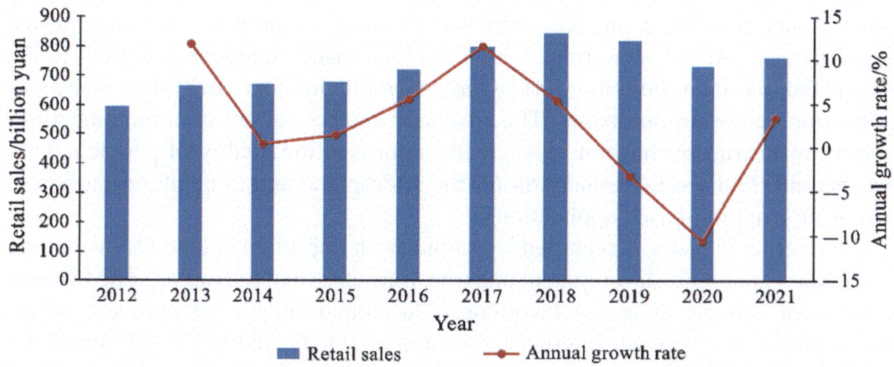

Fig. 4.32 China's home appliance retail sales and annual growth rate (2012–2021). *Data Source* All View Cloud

the 2020 "Global Smart Home Invention Patent Rankings" with 2034 patents, holding the global first place for the fourth consecutive time. In 2020, Midea had around 200 subsidiaries, 28 R&D centers, and 34 major production bases worldwide, including 18 overseas R&D centers, employing over 15,000 R&D personnel. By the end of 2020, Midea's cumulative patent applications exceeded 160,000, with over 6000 granted patents.

4.5 Summary

In the era of the booming digital economy, the traditional ICT industry is accelerating its transformation and upgrading, deeply integrating into the digital economy wave, and gradually developing into a key industry covering new digital products and cutting-edge digital technologies. China uses the concept of "digital industrialization" to summarize the development connotation of the ICT industry in the new situation of the digital economy. It involves multiple fields, including fundamental telecommunications industry, the Internet industry, software and information technology services industry, and electronic information manufacturing industry.

In the field of fundamental telecommunications industry, China gives full play to the guiding role of policies and the demonstration and leading role of state-owned enterprises, and continuously promotes the optimization and upgrading of the national communications network infrastructure. In the construction of Internet infrastructure, China has built the world's largest optical fiber broadband network, ranking first in the world in the number of IPv6 addresses and active users. In the field of mobile communications, China took the lead in breaking through 5G technology and built the world's largest 5G mobile network. The number of 5G base stations in China accounts for more than 60% of the global total. The construction of the satellite Internet is advancing steadily. Many types of satellites have been successfully launched and networked, and low-orbit satellite network projects are being carried out as planned. At the same time, China has also made significant progress in the IoT and the IoV. It has built the world's largest mobile IoT and established many IoV application demonstration bases. The advanced and cost-effective communications network infrastructure has not only greatly improved the quality of people's daily lives but also laid a solid foundation for the widespread and in-depth promotion of various information-based applications.

The Internet industry is regarded by China as an important engine for promoting high-quality economic development and transformation and upgrading. The Chinese government actively creates a favorable policy environment, encourages various market players to carry out Internet innovation activities, and vigorously promotes the in-depth application of the "Internet+" strategy in various fields. At the level of Internet applications, China shows strong competitiveness in fields such as e-commerce, social entertainment, and fintech. China's online retail value has ranked first in the world for many years. The instant messaging has a large user base, with over one billion monthly active users. The short-video industry is developing rapidly, with a user scale of more than one billion. Short-video platforms have become a new and important channel for digital marketing. In addition, China leads the world in online payment application scale. Mobile payment, with its convenience and efficiency, has profoundly changed people's payment habits. Alipay actively expands its global business and provides high-quality financial services to over 1.2 billion users worldwide. Relying on the booming Internet industry, new business forms and models such as the sharing economy, short-video platforms, and live-streaming e-commerce have emerged in an endless stream, strongly promoting the prosperous development of China's real economy.

As a strategic emerging industry, the software and information technology services industry has received high attention and strong support from the Chinese government. Through a series of industrial policies, China fully promotes the independent and controllable development of this industry and continuously improves the software innovation system. In the field of basic software, China has made a series of key breakthroughs, successfully developing many basic software products with independent intellectual property rights. These products cover areas such as server, personal computer, mobile terminal operating systems, as well as distributed databases, and have achieved large-scale applications in many industries. In emerging information technology fields such as cloud computing, big data, AI, and blockchain, Chinese enterprises are actively innovating and have launched a series of advanced technologies and products. The cloud computing market maintains rapid growth, and the competitiveness of public cloud services in the international market is constantly increasing. The big data industry is booming, and China's data sharing level ranks among the top in the world. The AI field has rich innovation achievements, and the industry scale continues to expand. Blockchain technology R&D is at the world-leading level and has already seen large-scale adoption in many fields such as finance. The rapid development of the new software and information technology services industry has injected continuous innovation vitality into the informatization construction of various industries.

As an important pillar of China's manufacturing industry, the electronic information manufacturing industry, with its complete industrial system, produces a large number of electronic devices and products globally. At the same time, China actively promotes the industrial upgrading of high-end manufacturing industries such as IC. In the semiconductor field, Chinese enterprises have faced difficulties and continuously overcome key technological "bottleneck" problems. They have made significant breakthroughs in semiconductor materials, equipment, IC, etc., and achieved the import substitution of most mid- to low-end products. In the field of new displays, Chinese enterprises have achieved a leap from following to leading in technology, built a complete industrial chain, and ranked among the top in the world in terms of industry scale and the number of technical patents. In the manufacturing fields of computers, communications equipment, and consumer electronics, China not only has strong manufacturing advantages but also continuously improves its technological innovation ability. China's electronic information products and technologies are widely exported around the world, providing global consumers with a wide variety of product choices.

In recent years, China has achieved remarkable development achievements in the fields of fundamental telecommunications industry, the Internet industry, software and information technology services industry, and electronic information manufacturing industry. The advanced network infrastructure has a wide coverage, building a high-speed channel for information circulation. Innovative applications in the Internet field emerge in an endless stream, shaping a convenient and efficient lifestyle and consumption pattern. The technological breakthroughs and product innovations in the software and information technology services industry provide strong support for the digital transformation of various industries. The industrial upgrading and technological progress in the electronic information manufacturing industry have enhanced China's position in the global industrial chain. These achievements have significantly improved China's social outlook: convenient network communications allow people to enjoy instant access to information anytime and anywhere; applications such as mobile payment and e-commerce simplify the consumption process; short-videos and instant messaging enrich people's social and entertainment lives. At the same time, these achievements have also played a crucial role in promoting China's overall economic development. They have given birth to a large number of new business forms and models, created numerous job opportunities, promoted the transformation and upgrading of the real economy, and driven the coordinated development of related industries, thereby serving as an important engine for promoting high-quality economic development in China, and serving an advantage for China in the global digital economy competition.

Chapter 5
Industrial Digitalization

Industrial digitalization refers to the application of information and communications technology (ICT) and data resources to increase output and improve efficiency in traditional industries. Over the past decade, China has actively promoted the digitalization process of various industries and facilitated industrial upgrading through digital transformation. This chapter focuses on the current status and achievements of China's digital development in industries such as manufacturing, agriculture, energy, and transportation.

5.1 Intelligent Manufacturing

China is a manufacturing powerhouse and has maintained the title of the "world's factory". From 2012 to 2021, the value-added by China's manufacturing industry increased from 16.98 trillion yuan to 31.4 trillion yuan, with its global share rising from 22.5% to nearly 30%. China consistently holds the position of the world's leading manufacturing nation. The number of industrial and information technology brands from China entering "The World's 500 Most Influential Brands" increased from 10 to 24 during this period.

China's manufacturing industry is comprehensive, comprising 31 major categories, 179 subcategories, and 609 minor categories, making it the most complete and integrated manufacturing sector globally. In May 2015, the State Council released the "Made in China 2025" plan, aiming to promote the integrated development of industrialization and informatization through the "Internet Plus" initiative. In April 2021, the Ministry of Industry and Information Technology (MIIT) issued the "14th Five-Year Plan for the Development of Intelligent Manufacturing", outlining goals of achieving comprehensive digitalization and networking for large-scale manufacturing enterprises by 2035, with key industry leaders achieving basic intelligence.

W. Liu et al. (eds.), *Digital China*, Introduction to China's S&T Innovation,
https://doi.org/10.1007/978-981-95-6240-4_5

5.1.1 Industrial Internet

Industrial Internet, as an emerging industry pattern and application model resulting from the deep integration of new-generation information technology and the manufacturing industry, is a crucial foundation for achieving the digital transformation of industries. With the convergence of ICT and the industrial sector, industrial equipment becomes increasingly networked, industrial software becomes more intelligent, and industrial manufacturing becomes more collaborative, leading to the emergence of various new applications in industrial manufacturing. In November 2017, the State Council launched the "Guidance on Deepening the Development of 'Internet Plus Advanced Manufacturing' to Promote the Development of the Industrial Internet". In December 2020, the MIIT issued the "Industrial Internet Innovation and Development Action Plan (2021-2023)", proposing 5 aspects, 11 key actions, and 10 major projects to address deep-seated issues in the development of the Industrial Internet.

The depth and scope of the Industrial Internet platform system continues to expand. Currently, China's comprehensive, distinctive, and professional Industrial Internet platform system is continuously improving. As of December 2021, China had over 150 influential Industrial Internet platforms and a total of over 76 million connected devices; benchmark comprehensive Industrial Internet platforms had significantly led the way, with 15 enterprises accelerating the construction of cross-industry, cross-domain platforms to support the development of industrial enterprises. Specialized Industrial Internet platforms focuses on in-depth industry cultivation, forming a differentiated development pattern. Professional Industrial Internet platforms are accelerating their development, covering specific areas such as digital twins, industrial intelligence, industrial big data analytics, edge computing, and remote monitoring.

In terms of the industrial Internet identification and resolution system, in June 2020, the China Academy of Information and Communications Technology (CAICT) was authorized by the International Association for Automatic Identification and Mobile Technology (AIM Global), and became an international code issuer alongside large international organizations such as the International Article Numbering Association, the Institute of Electrical and Electronics Engineers, and the Universal Postal Union. It was assigned the code "VAA" and has the capability to issue global identification codes. In January 2021, the Data Identifier Management Committee (DIMC), the sole global management organization of the data identifier list, formally approved the exclusive international data identifier "15N" for the Industrial Internet identification and granted it to the CAICT for management and maintenance, which has a significant impact on the international promotion of the Industrial Internet identification.

Efforts have been made to implement the digital transformation of the manufacturing industry and the intelligent manufacturing project, creating an upgraded version of the integration standardization of industrialization and informatization. As of October 2022, over 57000 enterprises in China had participated in the integrated management system standardization, with over 1500 pilot demonstration projects in digitalized, networked, and intelligent manufacturing; the number of industrial apps had reached 283200; over 300 national standards and 42 international standards for intelligent manufacturing had been released; nearly 100 suppliers with prime operating revenue exceeding one billion yuan had been nurtured, covering over 90% of manufacturing sectors such as automobiles, textiles, and pharmaceuticals.

5.1.2 Industrial Robot

The robotics industry in China is experiencing vigorous growth, injecting robust momentum into economic and social development. During the 13th Five-Year Plan period, China's robotics industry has demonstrated a positive trend through continuous innovation and deepened applications. The industry's scale has rapidly expanded, with a compound annual growth rate of approximately 15%. In 2020, the operating revenue of the robotics industry surpassed 100 billion yuan, production of industrial robots reached 212000 units; and the robot density in the manufacturing industry reached 246 units per 10000 employees, nearly double the global average.

The core competitiveness of the robotics industry continues to improve. China has overcome some technical challenges in core components such as reducers, controllers, servo systems, and the core components have been gradually localized. Taking the reducer as an example, the Y series harmonic reducer made by Leader Harmonious Drive Systems Co., Ltd., based on the principle of triple harmonic reduction, has significantly enhanced torsional rigidity and transmission accuracy. The application of machine vision technology has significantly improved the degree of flexibility and automation in industrial production. For example, Zhongke Xinsong Co., Ltd. combines machine vision with collaborative robots to provide stable and continuous three dimensions (3D) vision flexible positioning for collaborative robot operations.

Industrial robots have penetrated into complex and precise scenes. They are more flexible after integrating the flexible force control characteristics, realizing the application of higher precision and stronger sensitivity, and their popularization and application in complex and precise scenes such as assembly, grinding, and riveting have been accelerated. Chongqing Huashu Robot Co., Ltd., targeting computer, communications, and consumer electronic products (3C products), has introduced precision processing robots equipped with independently developed high-performance servo motors and control technology. These robots focus on breaking through typical processes in the entire production of notebook computers, achieving full-process robot production, and applying a complete set of automated factory systems.

5.1.3 Intelligent Factory

5.1.3.1 Electronics Manufacturing

Realizing flexible production and manufacturing to improve production capacity. Huawei, in cooperation with China Mobile, has realized flexible production and manufacturing using 5G technology at the Huawei South Factory in Songshan Lake, Guangdong Province (see Fig. 5.1). The original mobile phone production workshop required 90000 meters of wiring, with an average of 186 devices per production line. The production line is upgraded and adjusted every six months with the update of new mobile phone models, and all the network wires in the workshop are rearranged, with each adjustment requiring a two-week work stoppage. Through the integration of 5G and the Industrial Internet, Huawei South Factory has realized wireless connection of the existing machines, reflow ovens and dispensing machines of the production line through the 5G network, and shortened the adjustment time of each production line from two weeks to two days.

Implementing on-site assisted assembly to accelerate assembly efficiency. Haier, in cooperation with China Mobile, utilized 5G technology to realize on-site assisted assembly of precision industrial equipment in Qingdao City, Shandong Province. Haier's Qingdao Washing Machine Factory built a connected factory based on 5G MEC (mobile edge computing) and carried out 5G remote-assisted assembly based on augmented reality (AR) glasses. Workers collect on-site videos of key industrial equipment by wearing AR glasses, and at the same time retrieve product installation instructions from the back-office system and push them to the AR glasses to realize efficient assembly. Workers can also contact remote experts through the 5G network to realize remote guidance.

Deploying machine vision for quality inspection to enhance automated detection. Gree, in collaboration with China Unicom, utilized 5G technology in Guangdong

Fig. 5.1 Huawei: Songshan Lake mobile phone production base. *Image Source* Huawei

Fig. 5.2 Gree: machine vision quality inspection application. *Image Source* Gree

Province to implement machine vision quality inspection. Gree established an industrial virtual private network in its assembly workshops to realize the integration of the production control network with the production management network. By utilizing 5G network, the content to be inspected is automatically photographed, and the photo and video streams can be uploaded to the machine vision quality inspection application deployed on the MEC platform; by leveraging graphics processing unit (GPU) arithmetic resources and data models, real-time comparison and analysis inspections can be conducted, realizing the automatic identification of equipment; the quality inspection system then performs the operation of separating defective products (see Fig. 5.2).

5.1.3.2 Equipment Manufacturing

Creating factory-area smart logistics to promote logistics development. Foton Motor, in cooperation with China Unicom, built a super truck factory in Zhucheng of Weifang City, Shandong Province, and realized factory-area smart logistics using 5G networks. In the scheduling link of inbound vehicles, Foton Motor developed an inbound synergy system integrating virtual electronic fences, automatic vehicle recognition, vehicle detection and other technologies, and used 5G technology to transfer real-time information such as the status of vehicle parking in the factory to a variety of intelligent display terminals and information systems, realizing paperless receipt of goods.

Collaborating on research and development (R&D) and design to accelerate the R&D process. Commercial Aircraft Corporation of China,Ltd. (COMAC) cooperated with China Unicom to carry out the construction of the project of "5G Plus Industrial Internet Enabling Intelligent Manufacturing of Large Aircraft" in Shanghai Pudong New Area. Based on 5G network services, COMAC supports cross-region, real-time online collaboration and remote diagnosis during the R&D and experimental phases of products through real-time uploading of AR data (see Fig. 5.3), which has

Fig. 5.3 COMAC: AR glasses assisted assembly remote diagnosis. *Image Source* Xinhua News Agency

realized issue identification in R&D and design and compressed the time cost of R&D experiments.

Facilitating collaborative equipment operations to enhance scheduling efficiency. Sany Heavy Industry, in cooperation with China Telecom and Huawei, carried out the construction of "5G plus Industrial Internet" private network project in Beijing, and 5G technology has been deeply integrated with the production process of machinery manufacturing to realize the cooperative equipment operation. The workshop network has been built through 5G technology, and the 3D images and status information of the automated guided vehicle (AGV) are transmitted based on the 5G network with large bandwidth and low latency; the 5G MEC platform and GPU arithmetic integration capability are utilized to reduce the complexity and cost of the single-machine function of the automatic guided vehicle and improve the efficiency of production scheduling.

5.1.3.3 Steel Production

Realizing remote equipment control and improving operational efficiency. Hunan Valin Steel Group Co., Ltd. (Valin Steel) cooperated with China Mobile to realize the remote control for overhead cranes and slag-adding robotic arms based on 5G technology in Hunan Province (see Fig. 5.4). By utilizing 5G's large uplink and download speed, operators can control the unloading of overhead cranes, lifting and loading slots, and overhauling operations in the remote control room in real time, ensuring the accuracy and real-time nature of remote control and improving operational efficiency.

Implementing machine vision quality inspection to improve monitoring efficiency. Anshan Iron and Steel Group Co., Ltd. (Ansteel) cooperated with China Mobile to carry out the constructing of the project of "R&D and Application of 5G-based Machine Vision Strip Steel Surface Inspection Platform" in Liaoning Province (see Fig. 5.5). The 5G network is deployed to transmit the collected high-definition image

Fig. 5.4 Valin Steel: remote control of overhead cranes and robotic arms. *Image Source* China News Network

data from the cold rolling site to the platform in the operation room, and then the platform's visual AI analysis capability is leveraged to process and analyze the images, realizing the real-time detection of surface defects on strip steel. After the deployment of the program, the detection rate of conventional defects in strip steel has been improved, reducing the downtime of strip breakage and injured roll change caused by defects in strip steel.

Conducting equipment fault diagnosis to enhance predictive accuracy. Baoshan Iron & Steel Co., Ltd. (Baosteel) cooperated with China Unicom to carry out the construction of the project of "5G Plus Industrial Internet High-quality Network and Public Service Platform for Process Industry" in Zhanjiang City, Guangdong Province, and realized the fault diagnosis of continuous casting rolls, fans, and other equipment by utilizing 5G technology (see Fig. 5.6). Through the 5G network, equipment data are transmitted in real time to relevant systems for equipment fault diagnosis. Utilizing artificial intelligence (AI) and big data technology, the project can predict equipment lifespans, reducing the workload associated with on-site wiring and improving the predictive accuracy of equipment lifespans.

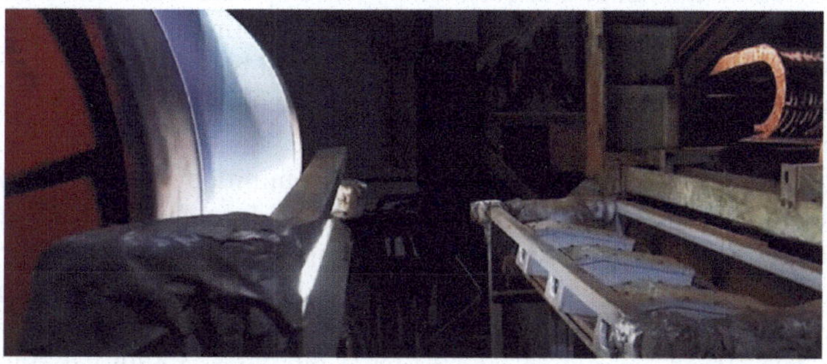

Fig. 5.5 Ansteel: machine vision strip steel surface defect detection. *Image Source Ansteel Daily*

Fig. 5.6 Baosteel: remote industrial control. *Image Source Zhanjiang Daily*

5.2 Smart Agriculture

China, as a prominent agricultural powerhouse, ranks among the world leaders in agricultural production. Despite this, the per capita agricultural resource of China is relatively scarce. At the end of 2019, China's arable land area was 1.918 billion mu, while the per capita arable land area was only 1.36 mu, which is less than 40% of the world average. At the same time, the spatial distribution of China's arable land resources is uneven, the overall quality is not high, and more than half of China's arable land relies on the natural conditions.

Since 1949, China has elevated its agricultural development through scientific breeding, water conservancy projects, and the widespread adoption of agricultural machinery, achieving self-sufficiency in the three staple foods. With less than 9% of the world's arable land, China produces approximately 25% of the world's grain output, sustaining nearly 20% of the global population and garnering remarkable achievements.

In recent years, China has implemented various policies to promote the modernization of agricultural science and technology, including the "14th Five-Year Plan for National Agricultural and Rural Informatization Development", the "13th Five-Year Plan for Agricultural Science and Technology Development", and the "Digital Agriculture and Rural Development Plan (2019-2025)". Starting from 2011, the Chinese government has allocated funds from the central fiscal budget to support the implementation of the National Internet of Things Application Demonstration Project's smart agriculture initiatives at the local level.

In February 2021, the "Opinions of the Central Committee of the Communist Party of China and the State Council on Comprehensively Advancing Rural Revitalization and Accelerating Agricultural and Rural Modernization" was issued, emphasizing the development of smart agriculture. Smart agriculture involves the deep integration of modern information technologies such as the Internet, the Internet of Things (IoT), big data, cloud computing, and artificial intelligence with agriculture. This integration facilitates a new agricultural production model characterized by agricultural information perception, quantitative decision-making, intelligent control, precise input, and personalized services. Smart agriculture relies on the support of information technology, utilizing high-tech solutions and scientific management to maximize resource efficiency.

5.2.1 Agricultural Big Data

China has successfully established a comprehensive agricultural geographic big data platform. The Ministry of Agriculture and Rural Affairs (MARA) has built platforms for geographic information public services, government data sharing, and agricultural and rural big data. The national three-dimensional "One Map" of natural resources continues to be improved, and results of delineations for arable land and permanent basic farmland, ecological protection redlines, and urban development boundaries (the "Three Lines") have been mapped and stored(see Fig. 5.7). Various business data layers related to agriculture and rural areas have been constructed, forming a "One Base Map + N Agricultural Special Application Maps" model. This enables hierarchical presentation, intuitive comparison, and dynamic tracking of data with fine granularity and visualization. A three-dimensional investigation system for new types of crops, covering "space, sky, and land", has been initially established. This system allows for the accurate acquisition of data on major crops sowing areas, spatial distribution, and crop growth conditions, etc.

Efforts have been made to enhance the early warning and monitoring of significant natural disasters in agriculture. The meteorological information warning and agricultural situation dispatch system played a crucial role in addressing the flood disaster during the 2021 autumn-winter planting season and the meteorological drought in the

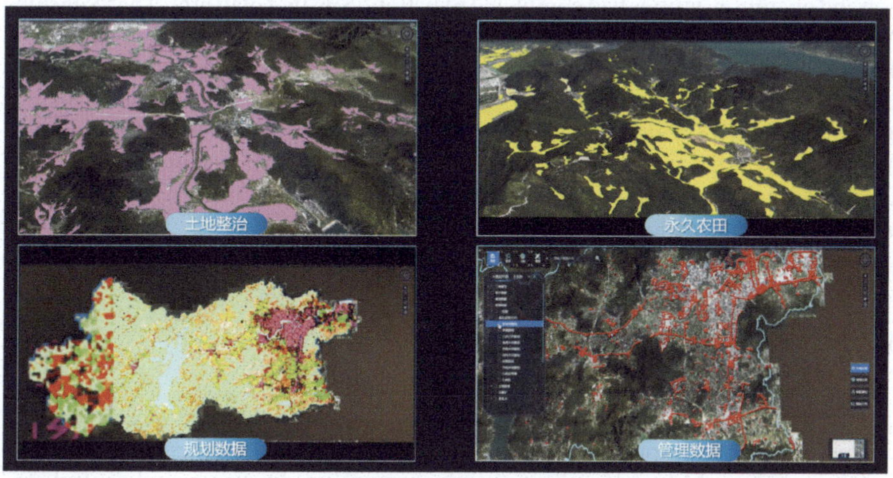

Fig. 5.7 Three-dimensional display of natural resources (land remediation, permanent farmland, planned data, etc.). *Image Source* Taizhou Bureau of Natural Resources and Planning

Yangtze River Basin in 2022. Monitoring of agricultural non-point source pollution and irrigation water usage has been comprehensively strengthened, with a total of 3882 monitoring points for agricultural non-point source pollution control sections as of June 2022. The national digital monitoring and early warning system for major crop diseases and pests has been continuously improved, with integration into 22 provincial platforms and more than 4000 IoT devices. This system provides robust support for the effective detection and control of major crop diseases and pests such as wheat stripe rust, rice brown planthopper, and fall armyworm.

China has initiated agricultural product market monitoring and early warning systems. A series of big data analysis application centers for the whole industrial chain of single agricultural products have been established. Efforts have been intensified in constructing an agricultural monitoring and early warning system, improving the analysis system for the balance between agricultural product supply and demand, and guiding various market entities to explore and innovate the application of agricultural and rural big data. Pilot projects for the whole industrial chain big data construction of 15 varieties in eight categories, including soybeans and apples, are steadily progressing. The information data platform for hog products has gone online, providing data for the whole pork industry chain. A data channel has been established on the MARA website, issuing the Agricultural Product Wholesale Price 200 Index. The analytical and judgment capabilities for the agricultural product market have significantly improved.

China has implemented traceability systems for the quality and safety of agricultural products and agricultural supplies. The National Agriculture Food Quality Safety Traceability Management Information Platform has achieved connectivity with 31 provincial platforms and agricultural reclamation platforms. The "Compliance and Qualification Certificate plus Traceability Code" model has been promoted, enabling full-process traceability from field to table. As of June 2022, 465000 production and operation entities had completed registration on that platform. The "Sunshine Agriculture Safety" pilot program was underway in five provinces, contributing to the continuous improvement of the agricultural product quality and safety traceability system. The informatization management of agricultural supplies such as pesticides, veterinary drugs, and fertilizers have been comprehensively promoted. As of August 2022, the Pesticide Digital Supervision & Management Platform of China had achieved 100% traceability with a "one bottle, one code" system for pesticide products nationwide.

5.2.2 Smart Cultivation

Guided by the MARA, modern ICT is rapidly applied in the field of crop cultivation, witnessing widespread adoption of technologies and equipment such as precision sowing, variable-rate fertilization, smart irrigation, environmental control, and agricultural unmanned aerial vehicle (UAV). By the end of 2021, China had established nine agricultural IoT demonstration provinces, implemented 100 digital agriculture pilot projects, and collected and released 426 agricultural IoT application achievements and models of reducing costs and increasing efficiency.

Agricultural UAV has gained extensive application in agriculture. Compared with manual labor or traditional machinery, UAV demonstrates significant advantages in tasks such as crop seeding (pollination), pesticide spraying, fertilization, as well as monitoring crop growth and detecting pests and diseases. China has become the world's largest user of agricultural UAV. In 2020, approximately 150 million mu of arable land in China utilized crop protection and remote sensing UAV operations, accounting for 8.3% of the country's arable land area, surpassing Japan's UAV-operated area by 100 times. According to data from the MARA, over 30000 UAVs were put into use during the spring planting season in 2021. As of 2022, the national inventory of crop protection UAV reached 121000 units, with an annual operation of 1.07 billion mu. Fig. 5.8 shows the application of a 5G network-connected crop protection UAV developed by Chongqing Academy of Agricultural Sciences.

Unmanned agricultural machinery and equipment have achieved widespread application. As of 2022, the accuracy of the BeiDou Navigation Satellite System (BDS) terminal for agricultural machinery has been improved to two meters; over 600000 tractors and combine harvesters had been equipped with BDS-based operation monitoring and intelligent control terminals, with more than 100000 tractors featuring auxiliary autopilot systems; the data platform had aggregated 20 billion agricultural machinery comprehensive data from 490000 agricultural machinery BDS terminals, enabling real-time collection and dynamic display of national agricultural machinery operation data. In 2021, the operation area served by BDS in China exceeded 60 million mu. Specifically, Heilongjiang Reclamation Area built six smart (unmanned) farm clusters, and had modified and upgraded a cumulative total of 6288 unmanned and assisted driving machinery in paddy and dry fields, with a demonstration operation area of 6.08 million mu, and an average yield increase of 3% to 5% per mu. Fig. 5.9 shows the scenes of BDS-assisted unmanned rice transplanter and unmanned harvester in Hongwei Farm of Beidahuang Group.

Fig. 5.8 5G network-connected crop protection UAV developed by Chongqing Academy of Agricultural Sciences. *Image Source* Red Star News(left),Xinhua net(right)

Fig. 5.9 BDS-assisted unmanned rice transplanter and unmanned harvester used in Hongwei Farm of Beidahuang Group. *Image Source Heilongjiang Daily*

The IoT-assisted precision farming has also been applied on a large scale. In the Smart "Wuhu Rice" Production Demonstration Base in Wuhu City, Anhui Province, the rice production process is divided into 13 stages, including sowing, transplanting, tillering, etc. Additionally, 49 smart decision points, such as variety selection, land leveling, and nitrogen fertilizer usage, are identified. The base has established a dual-drive technology system with "smart agronomy + intelligent agricultural machinery", enabling information perception, quantitative decision-making, and intelligent operations throughout the entire cultivation cycle (see Fig. 5.10). In 2022, the test area expanded to 150000 mu, and the test results showed an average annual increase of 14.3% per mu, nitrogen fertilizer savings of 32.5%, phosphorus fertilizer savings of 16.8%, medicine reduction of 38.0%, and an average income increase of about 500 yuan per mu. According to statistics, in 2021, the informatization rate of China's field cultivation was 21.8%, of which the production informatization rates of wheat, rice, cotton and corn were 39.6%, 37.7%, 36.3% and 26.9%, respectively.

Fig. 5.10 Paddy field sensors and its automatic irrigation system used in the Smart "Wuhu Rice" Production Demonstration Base in Wuhu City, Anhui Province. *Image Sources Wuhu News* (left), Wanjiang Pearl Network (right)

5.2.3 Smart Farming

Modern ICT has been widely used in the whole process of livestock and poultry farming. In 2021, the informatization rate of livestock and poultry farming in China reached 34.0%, with the informatization rates for hog and poultry farming being 36.9% and 36.4%, respectively. Comprehensive information platforms for animal husbandry and feed and fresh milk quality supervision systems had achieved comprehensive supervision over more than 180000 large-scale pig farms, more than 4200 fresh milk acquisition stations, over 5800 transportation vehicles, 300 or so ranches, and approximately 13000 feed production enterprises holding licenses. Various types of farms had increased labor productivity by over 30% through the application of unmanned environmental control platforms, automatic inspection and alarm systems, intelligent feeding systems, etc., reducing costs by approximately 150 yuan per head of slaughtered hog.

Smart farms has realized intelligent monitoring of farming environments and individual behaviors of livestock and poultry. For instance, in the fully automated unmanned "future pig farm", 5G-enabled mobile patrol robots scan pig pens back and forth, displaying real-time data such as pig body temperature and pen temperature on a large screen. The "future pig farm" has achieved intelligent, digital, and unmanned management throughout the entire process of hog farming. The average annual production of market pigs per sow increased from 18.5 heads in conventional farming methods to 26.5 heads, achieving an efficiency increase of 43.2%. The labor requirement for 2500 hogs decreases from 4 people to 0.8 people, saving 80% of labor. The total water consumption of the farm has been reduced by 60%. Fig. 5.11 shows the "future pig farm" in Shimen Town, Tongxiang City, Zhejiang Province.

The fishery industry has been promoting factory farming, rice-shrimp farming, and aquaponics models based on digital technology. In 2021, the informatization rate of China's aquaculture was 16.6%, of which the production informatization rates of crabs, shrimps, fishes, and shellfishes were 23.6%, 21.6%, 20.9%, and 6.0%, respectively. Coastal provinces have been continuously carrying out the deployment of equipment such as BeiDou and Tiantong satellite terminals for marine fishing vessels. Leveraging the dynamic monitoring management system for fishing vessels,

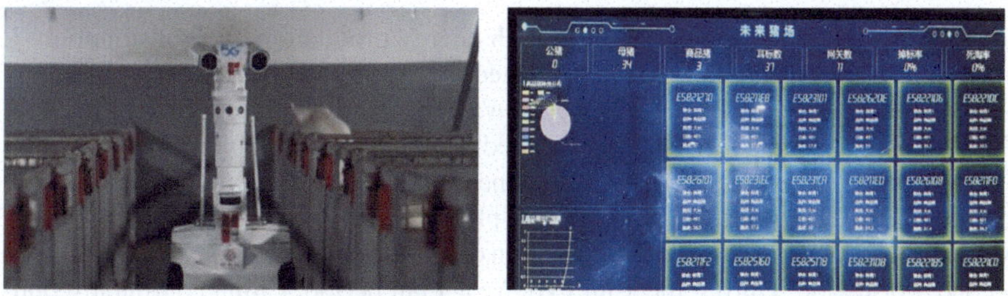

Fig. 5.11 The "future pig farm" in Shimen Town, Tongxiang City, Zhejiang Province. *Image Source* Guangming Net

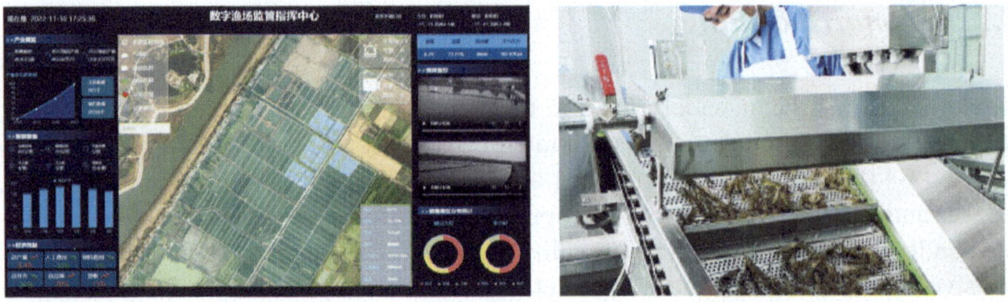

Fig. 5.12 The digital fishery in Pukou District, Nanjing City, Jiangsu Province. *Image Sources* njdaily.cn (left), Media Center of Pukou District (right)

China has built a comprehensive "one map" of national marine fishing vessel dynamic positions, creating a complete database for "fishing vessels + ports + crew, nearshore + offshore" fishing activities, providing strong support for off-season fishing bans and fishing vessel supervision.

In Pukou District, Nanjing City, Jiangsu Province, the entire process from production to distribution, and then to consumption of freshwater shrimp has been digitized to create an industry system based on order production, transparent supply, and trusted consumption. A digital freshwater shrimp farming system has been established to achieve online perception of farming environments, real-time regulation of water dissolved oxygen, and precise feeding. This has improved water quality, increased stocking density from 100000 to 120000 per mu, raised production by 20%, reduced labor input by 150 yuan per mu, and saved 200 yuan per mu in inputs. The survival time of freshwater shrimp during storage and transportation has increased from 1 hour to 10 hours, with a survival rate of over 98%. The entire system has reduced more than 15% of labor costs in farming, saved more than 20% in storage, processing, and logistics distribution costs, and increased farming income by more than 15%. Fig. 5.12 shows the digital fishery in Pukou District, Nanjing City, Jiangsu Province.

5.3 Smart Energy

China stands as a significant player in both energy production and consumption. It has fundamentally established a multi-faceted energy production system propelled by coal, oil, gas, electricity, nuclear power, new energy, and renewable energy. In 2019, China's primary energy production reached a substantial 3.97 billion tons of standard coal, making it the world's leading energy producer. The cumulative installed capacity of hydropower, wind power, and photovoltaic power ranked first globally, while the capacity of nuclear power plants in operation and under construction ranked second worldwide. During the 13th Five-Year Plan period, China continually optimized its energy structure, achieving remarkable success in the low-carbon transformation. The proportion of coal consumption decreased to 56.8%, with installed capacities for conventional hydropower, wind power, solar power, and nuclear power reaching 340 million kilowatts, 280 million kilowatts, 250 million kilowatts, and 50 million kilowatts, respectively.

In response to environmental challenges such as global warming, the world's 178 contracting parties signed the Paris Agreement in 2016. On September 22, 2020, Chinese President Xi Jinping declared at the General Debate of the 75th United Nations General Assembly that China would enhance its nationally determined contributions, adopt more robust policies and measures, and strive to peak its carbon dioxide emissions before 2030 and achieve carbon neutrality by 2060. Driving the digitalization and intelligence of the energy industry has been regarded as a pivotal strategy to guarantee China's carbon peaking and neutrality goals. In 2020, China introduced the new infrastructure initiative, encompassing ultra high-voltage power transmission lines, new energy charging stations, and other energy infrastructure, and emphasized the establishment of smart energy infrastructure. The "14th Five-Year Plan for Modern Energy System", released by the National Development and Reform Commission (NDRC) and the National Energy Administration (NEA) in January 2022, advocates for the digitalization of energy infrastructure, promoting the digital transformation of the energy industry and enhancing the application of new information technologies, including next-generation ICT, AI, cloud computing, blockchain, the IoT, and big data, in the field of energy.

5.3.1 Intelligent Coal Mine

In February 2020, eight ministries of the Chinese government jointly issued the "Guiding Opinions on Accelerating the Intelligent Development of Coal Mines", which emphasized the need to leverage ICT to establish intelligent systems with comprehensive perception, real-time interconnection, analytical decision-making, autonomous learning, dynamic prediction, and collaborative control. The objective was to achieve the intelligent operation of various processes in coal mining, including exploration, excavation, transportation, ventilation, washing and selection, safety assurance, and business management, in order to enhance coal mine safety production levels and ensure stable coal supply. By 2035, it envisioned the comprehensive implementation of intelligence across various coal mines, constructing integrated intelligent systems across multiple industry chains and systems, ultimately establishing intelligent coal mine systems characterized by intelligent perception, intelligent decision-making, and automatic execution.

The level of information infrastructure in the coal industry has significantly advanced. Operational networks in factory areas has transitioned from hundred-megabit and gigabit industrial Ethernet Ring to networks exceeding 10 gigabits, even reaching 100000 gigabits. Underground wireless communications systems has evolved from the deployment of 3G and 4G networks to the integration of 5G, Wi-Fi 6, F5G (fifth-generation fixed network), and other converged network communications.

Personnel positioning systems underground has upgraded from radio-frequency identification (RFID) interval positioning to ultra-wideband sub-meter precision positioning. In June 2020, the first 5G intelligent coal mine in the country was completed at Shanxi Xinyuan Coal Mine. In 2023, over one hundred coal mines nationwide had implemented 5G networking and applications, continuously deepening applications such as remote control of coal mining and driving face, panoramic video stitching, open-pit coal mine autonomous driving, AR intelligent inspections, and remote diagnostic.

Autonomous driving technology has been deployed in open-pit coal mines. The Inner Mongolia Dayan Coal Industry Co.,Ltd. conducted the "Demonstration Project of 5G plus Autonomous Trucks in Open-pit Coal Mines under Extremely Cold and Complex Climate Environment". The operating speed reached an industry-leading 40 kilometers per hour, with an autonomous system availability exceeding 96.7%, and the comprehensive efficiency of autonomous transport not lower than that of manned driving. This project marked the first 5G network unmanned driving project in China's open-pit coal mines, the first project for unmanned transformation of over 200-ton-level vehicles, and the first project to achieve all-weather three-shift unmanned formation operation. As of the end of 2021, China had 146 unmanned driving vehicles in open-pit coal mines. Fig. 5.13 shows the "unmanned mining truck" of the Inner Mongolia Dayan Coal Industry Co.,Ltd.

Progress has been made in the connected transformation of mining equipment. In 2021, the China Energy Investment Co.,Ltd. and Huawei released MineHarmony, the first Industrial Internet operating system in the mining sector, in Beijing. As of November 2022, more than 3300 sets of equipment had been deployed in 13 coal mines and 1 coal washing plant, and the applications supported include intelligent control of equipment, unmanned inspection of fixed places, and online upgrading of equipment. The MineHarmony system utilized 5G+AI video splicing technology to achieve remote control; it also utilized 5G to send back images of the main transport belt in real time, with AI algorithms identifying anomalies such as lumps of coal and anchors, so as to achieve full-time monitoring, which could reduce the number of inspection personnel in the underground by 20%. Fig. 5.14 shows the scene of MineHarmony system assisting coal mines to realize intelligent inspection.

Fig. 5.13 The "unmanned mining truck" of the Inner Mongolia Dayan Coal Industry Co.,Ltd. *Image Source* ccoalnews.com

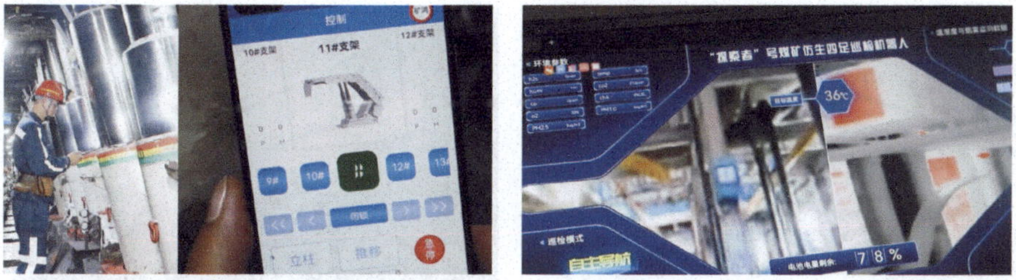

Fig. 5.14 Huawei's Industrial Internet operating system "MineHarmony" assists coal mines in realizing intelligent inspection. *Image Source* Wisdom Valley Trend

5.3.2 Smart Power Plant

In March 2023, the "Opinions of the National Energy Administration on Accelerating the Digitalization and Intelligence Development of Energy" was released, outlining the acceleration of clean and low-carbon transformation in power generation through the application of digital and intelligent technologies. The document emphasized expediting the digital design, construction, and intelligent upgrading of traditional power sources such as thermal power and hydropower, along with promoting the development and application of intelligent decentralized control systems. The concept of the "smart power plant" was introduced, where a new type of power plant is formed by integrating various systems with information technology, intelligent control technology, and power generation industry technology, based on the physical power plant.

Significant progress has been achieved in the intelligent transformation of thermal power plants. Traditional coal-fired thermal power remains a cornerstone of power generation in China, contributing approximately 71.13% to the country's total electricity generation in 2021, with a year-on-year growth of 8.4%. The Guodian Inner Mongolia Dongsheng Co-generation Power Co., Ltd. deployed a coal-piling robot in July 2019. The 3D laser unmanned coal-piling robot, equipped with a flexible guide rail system on the coal bin ceiling, created a 3D model of the coal pile in the coal yard. This achieved unmanned, intelligent patrolling through 3D laser unmanned coal-piling, requiring only five minutes for inventorying an entire coal yard. In 2020, the plant established a plant-level 5G network, achieving downstream rates of 350 Mbps, upstream rates of 160 Mbps, and bidirectional latency of less than 15 ms within the thermal power plant area, utilizing unmanned robots to enable comprehensive coverage of production and high-risk sites, providing a versatile inspection operation with multiple business and operation mode functionalities. Fig. 5.15 shows the "intelligent thermal power plant" of Dongsheng Co-generation Power.

Substantial progress has been made in the digitization of offshore wind farms. Typically located 20-120 kilometers offshore, offshore wind farms cannot be adequately covered by onshore mobile communications base stations. Previous communications in offshore areas primarily relied on satellite or narrowband private network communications, supporting only basic voice and simple data collection

Fig. 5.15 Dongsheng Co-generation Power's "intelligent thermal power plant". *Image Sources* National Energy Investment Group (top), National Energy Voice (bottom)

business requirements. In December 2022, State Power Investment Corporation Limited's Jieyang Shenquan II offshore wind power project in Jieyang, Guangdong Province, smoothly achieved full-capacity grid connection, boasting the world's largest single-unit capacity in a mass-applied offshore wind power project, with a total installed capacity of 502 MW. In collaboration with Huawei, China Mobile (Guangdong Province) deployed and activated two 5G base stations at the offshore booster station, 25 nautical miles from the coastline. This enabled network coverage for an area exceeding 1000 square kilometers, approximately 20-50 kilometers offshore, with download speeds exceeding 100 Mbps and network latency below 40 ms.5G technology facilitated high-definition video surveillance, offshore inspection, intelligent dispatching, and other applications in offshore wind farms. Fig. 5.16 shows a 5G offshore wind farm in Jieyang City, Guangdong Province.

5.3.3 Smart Power Grid

There is a significant geographical imbalance between China's energy resources and demand. Over 80% of energy resources are concentrated in the western and northern regions, while over 70% of energy consumption is concentrated in the

Fig. 5.16 5G offshore wind farm in Jieyang City, Guangdong Province, China. *Image Sources China Daily* (left), Huawei (right)

eastern and central regions, which necessitates large-scale, long-distance, and efficient electricity transmission. Long-distance and cross-regional electricity transmission requires precise regulation of the electric grid. The strategic goal of a "strong smart grid" was introduced in May 2009, with an emphasis on using ultra-high-voltage (UHV) grids as the backbone, coordinating the development of various levels of grids, and featuring characteristics of informatization, automation, and interaction. By the end of 2017, all planned UHV projects, known as "four crosses and four straights", were fully operational. These UHV projects, generally originating in resource-rich western and northern regions and terminating in the central and eastern regions, successfully met the growing electricity demand in load centers. By the end of 2020, China had essentially completed the construction of a strong smart grid, with technology and equipment reaching international advanced levels.

In March 2023, the "Opinions of National Energy Administration on Accelerating the Digitalization and Intelligence Development of Energy" was released, advocating for the use of digital and intelligent technologies to support the construction of new power systems. The document encouraged the digital presentation, simulation, and decision-making of physical power grids, explored the application of AI and digital twins in intelligent decision support and control of power grids, and enhanced the intelligent level of multi-energy complementary joint scheduling in the power system. Currently, China is accelerating the transformation from digitalizing the power grid towards evolving into an Internet of energy.

Visualization is a key aspect of planning and construction. The State Grid Corporation of China (State Grid) focuses on key areas such as production, operation, and service, melds various business systems and data, and realizes diagnosis of grid problems on the map, intelligent generation of planning, online review of projects, intelligent preference of plans, and automatic generation of reports. As of 2022, the production domain digitalization of China Southern Power Grid covered 933 application links in 68 application scenarios for four specialties of power generation, transmission, substation, and distribution (see Fig. 5.17). In terms of power transmission, a 3D digital twin covering over 1.3 million kilometers had been constructed, achieving over 250000 kilometers of autonomous flight routes and accumulating an AI training sample library of over 400000 graphics cards, reducing defect analysis time from 15-30 days to a few hours. In terms of power transmission, remote

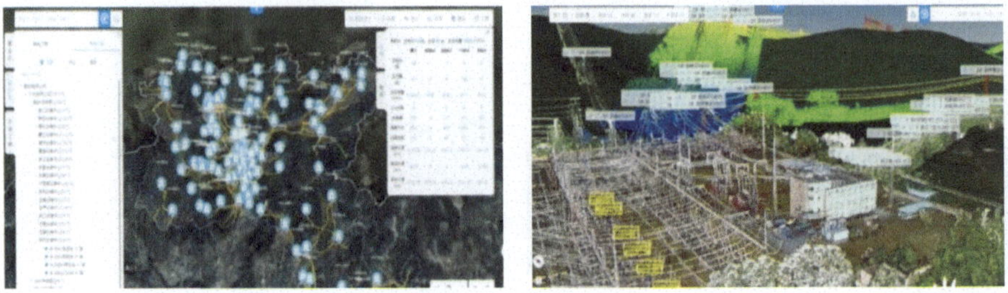

Fig. 5.17 China Southern Power Grid: digital twin of the main 110 kV and above transmission network. *Image Source* China Southern Power Grid

automatic control of substations had been implemented, achieving "unmanned and remote operation" of substations and improving the operating efficiency of substation equipment by 30%.

Production and operation have become increasingly intelligent. The State Grid has strengthened the digital control of grid scheduling and operation, and production and inspection. A new-generation scheduling technical support systems have been developed to achieve holographic perception of power grid operation status, observable and controllable generation-load resources, and efficient operation of scheduling tasks, supporting the safe operation of large power grids. In the field of operation and inspection, UAV for autonomous inspections have been widely promoted, enabling automatic identification of equipment inspection images, some regions have employed UAV with fire-extinguishing capabilities to clear foreign objects from power lines; intelligent inspections of substations (converter stations) have reduced the average inspection time in operational UAV substations (converter stations) from 120 minutes for human inspections to 40 minutes for machine inspections, forming a "machine-centric, human-assisted" inspection model (see Fig. 5.18).

The autonomy of power equipment is emphasized. Power chips provide the underlying core support for the development of the power industry and are the most crucial components in power control systems. However, the domestication rate of core chips in equipment such as power transmission, transformation, and distribution remains

Fig. 5.18 State Grid: UAV and unmanned vehicle involved in facility operation and inspection. *Image Sources* The Paper(left), State Grid(right)

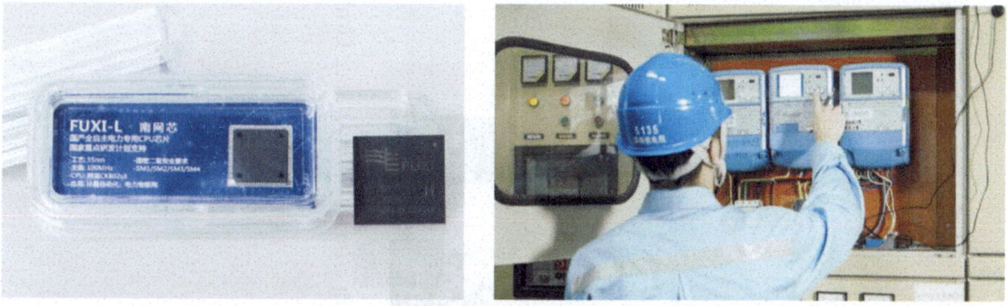

Fig. 5.19 China Southern Power Grid deploys domestic power specialized master control chip "Fuxi" in power grids. *Image Sources People's Daily* (left), *Yangcheng Evening News* (right)

relatively low. Since 2013, the level of domestic production of power hardware-level chips in China has been continuously improving. By the end of 2019, the "Fuxi" chip sample developed by the China Southern Power Grid was successfully delivered. It is based entirely on domestic instruction set architecture and domestic core, with comprehensive performance 1.5 times that of similar imported products. The "Fuxi" chip has been successfully applied in nearly 30 categories of devices, including power transmission, power distribution, new energy, and edge computing (see Fig. 5.19). In 2021, Shenzhen deployed 340 pieces of power equipment across eight categories, including charging piles, energy meters, and concentrators, with the "Fuxi" chip.

Digitalization of electricity services has been promoted. The State Grid has built the world's largest new energy cloud platform, providing one-stop services for new energy planning and construction, trading and settlement, etc. It has also built the world's leading smart Internet of Vehicles (IoV) platform with the widest coverage, the largest number of accessed charging piles, and the collaborative development of vehicles, changing piles, and power grid, providing convenient and intelligent charging and exchanging services for the travel of new energy vehicle owners. Furthermore, it has built one-stop services for household photovoltaic (PV) power station construction, grid connection, and settlement in the whole process, relying on the online platforms such as the "State Grid Online" APP and the new energy cloud, thus enhancing its ability to serve customers (see Fig. 5.20).

5.4 Intelligent Transportation

Since 1949, China has consistently prioritized and considered transportation infrastructure as the cornerstone and focal point of national economic development. It has constructed the world's largest transportation infrastructure. By the end of 2021, the total mileage of roads in China had reached 5.28 million kilometers, with the national expressway[1] spanning 117000 kilometers, ranking first globally. As of July

[1] Expressway refers to the multi-lane highways exclusively designed for vehicles to travel in separate directions and lanes, with fully controlled access. The speed limit is typically 60-120 km/h in China.

Fig. 5.20 "State Grid Online" APP allows users to handle services such as household PV power generation and new energy meters. *Image Sources* State Grid (left), Ningbo Publishing (center), *Zhejiang News* (right)

2022, the highway network, mainly composed of national expressways, has covered 98.8% of urban areas, cities with populations exceeding 200000, and prefecture-level administrative centers, connecting approximately 88% of county-level administrative regions and about 95% of the population. By the end of 2022, the total operating mileage of the national railway reached 155000 kilometers, with high-speed railway[2] covering 42000 kilometers, ranking first globally in high-speed railway mileage; a total of 53 cities had implemented urban rail transit projects, accumulating an operational mileage of 9584 kilometers, with subways accounting for more than 8000 kilometers, ranking first in the world; inland waterways had a navigable distance of 128000 kilometers. In 2022, China's waterways completed a freight volume of about 8.55 billion tons, with the Yangtze River mainline exceeding 3 billion tons, maintaining its position as the world's leading inland water transportation for over a decade. China's ports handled a cargo throughput of 15.68 billion tons, with Chinese ports occupying 8 out of the top 10 global ports and 7 out of the top 10 container ports in the world.

Simultaneously, the informatization of China's transportation infrastructure has advanced rapidly. According to the "Digital China Development Report (2020)", by the end of 2021, 21 provinces, autonomous regions, and municipalities had implemented the construction of highway and waterway and transportation market credit information service systems, achieving interconnected operation of the nationwide expressway electronic toll collection (ETC) system; 42 vessel traffic management systems (VTMS) and the automatic identification system (AIS) covering coastal areas, the Yangtze River mainline, and other inland waterways had been established; interconnected and interoperable transportation smart cards had been realized in 303 cities at or above the prefecture level; urban rail transit vigorously promoted smart subway services, serving a population exceeding 390 million with a daily average passenger volume exceeding 45 million nationwide; the "one-click ride-hailing" function of ride-hailing services had covered nearly 300 cities, and over 200 ride-hailing platforms had been connected to the online ride-hailing supervision information exchange platform, with a daily average order volume reaching 21 million.

In December 2021, the Ministry of Transport (MOT) issued the "14th Five-Year Plan for Digital Transportation", proposing six major development goals to be achieved by 2025: digital perception of transportation infrastructure, extensive coverage of information networks, convenient and intelligent transport services, online collaboration in industry governance, vibrant innovation in technology application, and robust cybersecurity guarantees. In December 2021, the State Council issued the"14th Five-Year Plan for the Development of the Modern Comprehensive Transportation System", proposing to build a modern, high-quality national comprehensive transportation network that is convenient, cost-effective, green, intelligent, and safe by 2035.

---

[2] High-speed railway refers to the closed rail transit system that supports the operation of high-speed trains and the operating speed is usually 250–350 km/h in China.

5.4.1 Smart Road Transportation

5.4.1.1 Road Facility Management

Roads constitute the fundamental component of China's transportation system, serving as the vital arteries for the healthy operation of the national economy. According to the "Digital China Development Report (2020)", the level of digitalization in China's transportation infrastructure had significantly increased, with widespread application of road construction information models and accelerated promotion of intelligent maintenance systems; all the highways had achieved comprehensive coverage of toll collection, communications, and monitoring systems, providing comprehensive support for expressway operations, management, and vehicle safety services; dynamic monitoring had been implemented for 50% of the key sections of national and provincial main roads, as well as special large bridges and exceptionally long tunnels. By the end of 2021, the density of China's road network had reached 55 kilometers per 100 square kilometers.

In terms of road facility management, in 2019, toll stations at provincial borders on national expressways were eliminated, and the ETC system was fully integrated across the country. By the end of 2020, the number of ETC lanes on expressways had reached 66000, with a total of 227 million ETC users; the utilization rate of ETC for passenger vehicles exceeded 70%, and for freight trucks, it surpassed 56%; the national expressway system had 28000 ETC gantries, 250000 sets of antennas and license plate recognition devices, accumulating over 1 petabyte of recorded data. In 2022, the Highway Monitoring and Response Center of the MOT issued the construction of the National High-speed Traffic Network's TraffiCatcher Integrated System. This system utilizes highway monitoring videos to detect traffic incidents, relay the information to the road network monitoring command center, and notify nearby ETC gantries and traffic information boards for timely traffic flow management.

In the realm of vehicle information management, the MOT leverages BDS to enhance the management of special vehicles. Since August 2011, newly manufactured "Two Passengers and One Hazardous" vehicles (where "Two Passengers" refer to tourist charter buses and scheduled passenger buses, and "One Hazardous" denotes vehicles transporting hazardous materials) are required to be equipped with BDS terminals before leaving the manufacturing factory. A nationwide intelligent monitoring platform for key "Two Passengers and One Hazardous" operational vehicles has been established, enabling dynamic monitoring and management of the operational status of key vehicles. By the end of 2021, more than 7.9 million operational vehicles had been installed with BDS terminals.

Taking Guangdong Province as an example, in February 2021, the "Two Passengers, One Hazardous, and One Heavy Cargo" Key Vehicle Intelligent Monitoring and Early Warning Convergence Platform (see Fig. 5.21) was formally put online. An intelligent video monitoring system was installed free of charge on 405000 key vehicles across the province. The system monitors the driver's driving behavior and the vehicle's trajectory (speed, geographic location, etc.), identifying and providing early warnings for various types of traffic hazards or violations of the law. After the deployment of this system, the number of traffic accidents and facilities in Guangdong Province in 2021 decreased by 62.87% and 59.26% respectively compared with the same period in 2020.

5.4.1.2 Intelligent Connected Vehicles

China has initiated research on autonomous driving since the 1980s, progressing through five levels (L1 to L5) based on the degree of automation. At L2, partial automation of functions, such as adaptive cruise control, lane-keeping assistance, automatic emergency braking, and automated parking, can be achieved under driver supervision. L4 represents highly autonomous driving, where the system handles all driving operations, and in specific scenarios, it requests the driver's response, allowing the driver to opt not to respond.

As of November 2022, China had witnessed sales of over 8 million vehicles equipped with L2 intelligent driving assistance functions, with a penetration rate of around 33%. The actual road test mileage for L4 autonomous driving had exceeded 40 million kilometers. In 2013, Baidu began embark on the development of autonomous vehicles. The autonomous vehicle solution of Baidu relies on the technologies of traffic scene object recognition and environmental perception, which can achieve high-precision detection, identification, tracking, distance and speed estimation, road segmentation, and lane detection. In December 2015, that solution achieved full autonomous driving in mixed road conditions, including urban streets, ring roads, and highways for the first time in China. In 2019, Baidu,in collaboration with the China First Automobile Group Co., Ltd. (short for FAW), produced China's first batch of mass-produced L4 autonomous passenger vehicles. Each vehicle was equipped with a LiDAR (LightLaser Detection and Ranging) sensor on the roof to sense the environment around it, with an effective detection range of up to 240 meters. In October 2020, Baidu's autonomous driving ride-hailing service was fully launched in Beijing.

The intelligent connected vehicle utilizes modern communications and network technologies, equipping the vehicle with sensors to perceive complex environments. If corresponding sensors and communications devices are installed on the roadside, the sharing of information between the vehicle and the infrastructure would be achieved. The onboard computing unit achieves information integration and intelligent decision-making, and the actuators execute actions according to the instructions issued by the computing unit, allowing the vehicle to travel safely in an efficiently energy-saving manner, realizing intelligent vehicle control.

In July 2021, the MIIT, the Ministry of Public Security (MPS), and the MOT jointly released the "Intelligent Connected Vehicle Road Test and Demonstration Application Management Regulations (Trial)". As of August 2021, China had established 16 intelligent connected vehicle test demonstration areas, providing access to over 3500 kilometers of test roads, and more than 700 test licenses were issued, accumulating a total test mileage of over 7 million kilometers. Taking Wuhan as an

Fig. 5.21 The application scenarios of "Two Passengers, One Hazardous and One Heavy Cargo" Key Vehicle Intelligent Monitoring and Early Warning Convergence Platform of Guangdong Province. *Image Sources Yangcheng Evening News* (left), Southern News Network (right)

Fig. 5.22 National Intelligent Connected Vehicle (Wuhan) Test Demonstration Zone. *Image Source Hubei Daily*

example, the National Intelligent Connected Vehicle (Wuhan) Test Demonstration Zone (see Fig. 5.22) was officially inaugurated in September 2019. The planned test roads cover 159 kilometers, encompassing residential areas, commercial areas, logistics zones, tourist scenic areas, and industrial zones. By adopting the "5G+BeiDou" vehicle-road cooperative network, the zone was developed into a national vehicle-to-everything (V2X) vehicle-road cooperative test area, achieving millisecond-level latency and centimeter-level positioning accuracy. Along the test roads, autonomous driving bus demonstration routes were launched and demonstration applications such as unmanned logistics delivery, unmanned sweeping, and smart parking were implemented.

5.4.2 Smart Rail Transit

5.4.2.1 Smart High-speed Railway

The Chinese government considers high-speed railway as a crucial national public transportation infrastructure and has actively pursued its development. Given China's vast territory, when the coastal regions got developed first during the reform and opening-up, the other regions of China had uneven economic development, particularly the less developed central, western, and northeastern regions. As a high-capacity, long-distance mode of transportation, high-speed railway serves to reduce both temporal and spatial distances, facilitating effective connections between resources dispersed across regions and accelerating the integration of logistics, passenger flow, information flow, and capital flow.

China has built a huge "four longitudinal and four transverse" high-speed railway network, with the mileage of the high-speed railway ranking first in the world, reaching about 95% of the cities with a population of more than 500000 people, and forming "hourly economic circles" around a number of major cities. In July 2016, the NDRC issued "Medium-and Long-Term Railway Network Plan", which explicitly proposed the construction of "eight longitudinal and eight transverse" high-speed railway main channels. It is expected to be completed by around 2026. In 2020, China proposed the new infrastructure initiative, which included high-speed railway and urban rail transit, aiming to establish the new intelligent transportation infrastructure. Alongside the development of China's high-speed railway network, significant progress has been made in the construction of rail supporting facilities and the intelligence of high-speed trains.

In terms of intelligent railway construction, the Beijing-Zhangjiakou high-speed railway is the first high-speed railway in the world designed with an intelligent approach (see Fig. 5.23). This project is geared toward the needs of the 2020 Beijing Winter Olympics, connecting Beijing and Zhangjiakou. The project started construction in April 2016 and opened for operation in December 2019, with a mainline length of 174 kilometers and a maximum design speed of 350 km/h. It was a large-scale and complex systematic project, involving multiple professional disciplines and presenting significant challenges in coordination. For this reason, it adopted new-generation information technologies such as building information modeling (BIM), big data, AI, BDS, 5G, and carried out intelligent innovation on the construction, equipment and operation technologies. The project implemented a BIM-based multi-disciplinary collaborative design platform and independently developed various professional BIM collaborative design software for surveying, routing, bridges, tunnels, roadbeds, catenary systems, signals, etc., saving approximately 8% of coordination liaison time and 3% of material costs.

In the realm of high-speed train intelligence, the Beijing-Zhangjiakou high-speed railway has achieved the sophistication of its multiple-unit trains. The new-generation

Fig. 5.23 Beijing-Zhangjiakou high-speed railway: intelligent construction of high-speed railway engineering. *Image Source* WANG TJ, The intelligent Beijing-Zhangjiakou high-speed railway, *Engineering*, 2021, 7 (12): 1665–1672

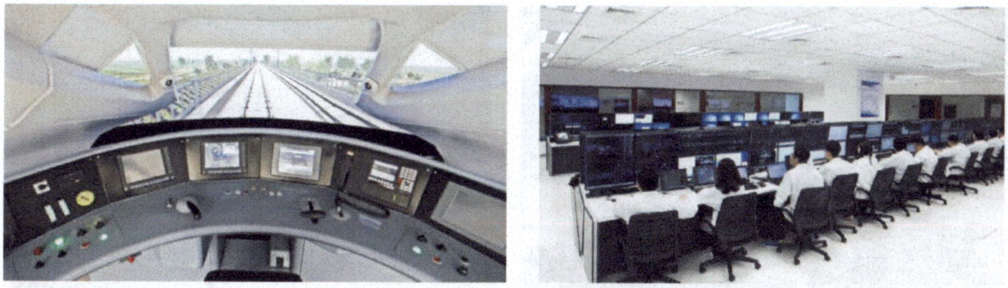

Fig. 5.24 Beijing-Zhangjiakou high-speed railway: automatic driving of high-speed trains. *Image Source* CCTV.com

bullet train system integrates a suite of technologies such as intelligent composite sensors, real-time transmission of train-ground information, big data mining and analysis, automation control, intelligent information processing, fault prediction, and health management. This integration enables intelligent train operations, intelligent maintenance (with monitoring points reaching up to 2718), intelligent services, and safety monitoring (introducing 168 new vibration monitoring points of the running gear). By adding automatic train operation (ATO) related equipment on the basis of the Chinese Train Control System, Level 3 (CTCS-3), the trains have achieved automatic driving (see Fig. 5.24). Additionally, precise positioning responders have been installed on the track in station areas, facilitating the trains automatic and precise alignment parking. This accomplishment encompasses automatic train departure at stations, automatic interval running, automatic stopping, automatic door opening, and synchronized door operation with platform doors. The Beijing-Zhangjiakou high-speed railway had set the world record for the first 350 km/h multiple-unit bullet train automatic driving.

5.4.2.2 Smart Metro

Along with the development of urbanization, China's urban population has been increasing, and the pressure on urban transportation has become increasingly significant. According to the results of the seventh national census in 2021, there were 21 megacities in China with a permanent resident population of more than five million, among which seven cities, including Beijing, Shanghai, and Guangzhou, have a population of more than 10 million. Cities with high population density in China are mainly concentrated in the Yangtze River Delta, Pearl River Delta, and Beijing-Tianjin region, with the population density of the top eight cities exceeding 2000 people per square kilometer. Metro is an important part of the comprehensive 3D transportation system of cities and an important means to alleviate the traffic pressure of large cities. According to the MOT, by the end of 2022, 53 cities in China had opened and operated 290 urban rail transit lines, with a total operating mileage of about 9584 kilometers. Along with the construction and development of metros,

Fig. 5.25 Beijing Metro: weather awareness and customer flow scheduling. *Image Source Beijing Youth Daily*

China has begun to explore the use of various types of information technology to improve the digitalization of metro operation and management.

Traditional metros are operated on a single line and managed manually. The smart station can break down the barriers between various specialties of the metro, realize the linkage between the equipment, and build an intelligent integrated control platform in the metro station. Taking the example of demonstration project by Beijing Metro, which was supported by the National Key Research and Development Plan "Key Technology for Efficient Transportation and Safe Service of Mega-City Rail Transit System", it relies on BDS, 5G, spatial digitization and other emerging technologies to build the Beijing Capital International Airport line into China's first full-scenario smart metro demonstration line. The system data in the station are connected to the meteorological data, which can receive real-time weather warning information, and the intelligent detection equipment in the station could monitor the rainfall and water level outside the station (see Fig. 5.25), and automatically activate the corresponding equipment when encountering dangerous situations, so as to solve the emergency situation on the spot. Through the verification of BeiDou positioning and a multi-mode navigation system, passengers can enjoy accurate location query, path planning, and other services on the Beijing Capital International Airport line through the Beijing Metro APP and WeChat mini-programs in the future, shortening the time of entering the station and improving the travel efficiency.

5.4.3 Smart Water Transportation

Compared with land transportation such as roads and railways, water transportation has the comparative advantages of high capacity, low cost, low energy consumption, and low pollution, and it is the main direction of China's efforts to accelerate the restructuring of transportation and promote the green and low-carbon transformation of transportation. As an important part of water transportation, ports and waterways are important hubs and channels of the national comprehensive transportation network.

China has established a water transportation infrastructure system with coastal shipping routes, the Yangtze River mainline, the West River shipping mainline, the Beijing-Hangzhou Grand Canal, and the Huai River main stream as the primary waterways. This system, centered around major ports, connects railways, highways, pipelines, and other modes of transportation, linking the world and forming a seamless connection between trunk and branch waterways. Water transportation handles a significant portion of cross-regional cargo transportation within China and is responsible for transporting about 95% of the country's foreign trade goods.

5.4.3.1 Digital Waterways

By the end of 2021, China's ports had 20867 berths for production terminals, 2659 berths of 10000 tons and above, and the navigable mileage of inland waterways was 127600 kilometers, of which the navigable mileage of Class III waterways and above was 14500 kilometers, and there were 125900 water transport vessels. The Yangtze River is the busiest waterway in China, and its freight volume has ranked first in the world for many consecutive years.

The construction of "Digital Yangtze River" has been advancing continuously, with initial gradually emerging. On September 30, 2019, the comprehensive connectivity of the digital navigation system along the Yangtze River officially commenced operation, enabling the dynamic monitoring of information such as navigation marks, water conditions, controlled river sections, and channel dimensions along the 2687.8 kilometers main channel of the Yangtze River. The establishment of the navigation maintenance and management platform, navigation dynamic monitoring platform, and navigation emergency command platform, with the mobile application Electronic Channel Map of Yangtze River and the Yangtze River Navigation Comprehensive Service Information System (portal website) as the carriers, had marked the initial success of the "One Map, One Station, Three Platforms" approach (see Fig. 5.26). The service model for managing the Yangtze River waterway is transitioning from a traditional manual mode to a digital mode, realizing the principles of "remote observation, precise management, and on-site utilization".

5.4.3.2 Smart Ports

China holds the distinction of being the country with the highest maritime connectivity globally, having established maritime routes with main ports in over 100 countries and regions. Its pivotal role in the global shipping and logistics system continues to rise. Among the top 10 ports worldwide in terms of cargo throughput and container throughput, China claims eight and seven positions, respectively. Shanghai Port, Ningbo-Zhoushan Port, and Shenzhen Port consistently ranked at the forefront in container throughput for multiple years. In 2021, ports in China handled a total cargo throughput of 15.55 billion tons, marking a 6.8% increase from 2020 and ensuring the smooth flow of "outbound and inbound" goods.

Fig. 5.26 Water condition and shipping monitoring of digital waterways along the Yangtze River mainline. *Image Sources* Changjiang Waterway Bureau (top), Yangtze River Waterway (bottom)

Port logistics informatization continues to deepen. After years of construction and development, the efficiency of loading and unloading operations at specialized coastal terminals and the container throughput along hundred-meter shorelines are both at the forefront globally. China has essentially formed a port logistics operation system characterized by "one cargo manifest, one information network". The application of blockchain-based electronic container release platforms in port navigation has accelerated. The average processing time for container documents has been compressed from two days to within four hours. Information exchange on container vessels has been achieved among 19 major ports in China, Japan, and South Korea. The international maritime container transportation has fully implemented electronic data interchange (EDI) and been incorporated into the international trade "single window" system.

Significant achievements have been made in the automation of container ports. Several fully automatic container terminals, including Xiamen Yuanhai and Qingdao

Fig. 5.27 Monitoring room and unmanned AGVs at the automatic terminal of Shanghai Yangshan Deepwater Port Phase IV Automated Terminal. *Image Sources Shangguan News* (left), Xinhua News Agency (right)

Qianwan, have been completed. Among them, the automatic terminals of Shanghai Yangshan Deepwater Port Phase IV (see Fig. 5.27) is the world's most intelligent container terminal. The Intelligent Terminal Operating System independently developed by Shanghai International Port (Group) Co., Ltd. served as its "brain"; while the Equipment Control System independently produced Shanghai Zhenhua Heavy Industries Co., Ltd. served as its "nervous system". Battery-driven automated guided vehicle (AGV) have achieved unmanned driving, automatic navigation, and path optimization. The more than 100 rail-mounted cranes and over 160 AGVs at the terminal of Shanghai Yangshan Deepwater Port Phase IV Automated Terminal can load and unload cargo 24 hours a day without human intervention, achieving millimeter-level hoisting accuracy and significantly improving port operational efficiency. Shanghai Yangshan Deepwater Port Phase IV Automated Terminal handled a daily throughput of 20823.25 ETUs, with a per capita labor productivity reaching 213% of that of traditional terminals.

5.5 Summary

For a long time, China has attached great importance to leveraging the driving effect of ICT on the development of traditional industries. It has successively put forward the development strategies such as "boosting industrialization with informatization" and "promoting the deep integration of informatization and industrialization". Since the 13th Five-Year Plan period, in line with national conditions and the development needs of the times, China has meticulously formulated scientific development plans and strategies, and actively promoted the digital transformation and upgrading of key industries such as manufacturing, agriculture, energy, and transportation.

In the field of manufacturing industry, as the "world's factory" with a complete industrial system, China occupies an important position in the global manufacturing landscape. At present, China is actively addressing the challenges of transforming into an environment-friendly high-end manufacturing industry, and vigorously promoting the implementation and application of new technologies such as the Industrial Internet, industrial robots, and 5G in manufacturing. In industries such as electronic device manufacturing, equipment manufacturing, and steel production, "smart factories" are emerging in an endless stream. These factories have effectively improved production efficiency and optimized product quality, providing a Chinese model for the transformation and upgrading of the global manufacturing industry.

In the field of agriculture, China has achieved self-sufficiency in staple food, effectively ensuring national food security. However, the relatively insufficient per capita agricultural resources remains a problem. In response to this challenge, China has established multiple big data systems for agriculture, realizing functions such as disaster early warning, market monitoring, and product traceability. In the field of cultivation, technologies such as unmanned aerial vehicles, unmanned agricultural machinery, and IoT are widely used. They can not only improve the operational efficiency of large-scale farms but also enhance the operational accuracy of small-scale farms. In the field of livestock and fishery sectors, the application of robot and IoT technologies has achieved intelligent monitoring and digital management, injecting new impetus into the development of agricultural modernization.

In the field of energy industry, China is the world's largest energy producer and consumer. It actively fulfills its responsibilities in addressing climate change and has announced its carbon peaking and carbon neutrality goals to the world. China is accelerating the digital and intelligent development of the energy industry. In the field of intelligent coal mining, the information infrastructure has been upgraded. Breakthroughs have been made in autonomous driving and equipment interconnection. The construction of smart power plants has promoted the intelligent transformation of fields such as thermal power and wind power. The construction of smart power grids has effectively alleviated the regional imbalance between energy supply and demand and has achieved excellent results in grid visualization, intelligent operation, and the localization of power equipment.

In the field of transportation industry, China has built the world's largest transportation infrastructure network. To further improve the quality and efficiency of transportation, China is vigorously developing intelligent transportation. In the field of intelligent road transportation, the digital management level of road facilities has been significantly improved, the ETC system has been fully popularized, and technologies such as intelligent connected vehicles and autonomous driving have been applied on a large scale, placing China in the global leading group. In the field of intelligent railway transportation, the intelligent construction of high-speed railways has made impressive achievements, and the digital operation management of metros is continuously advancing. In the field of intelligent water transportation, the construction of digital waterways and the intelligent development of ports have achieved remarkable results, improving water transportation efficiency and service quality.

The digital transformation and upgrading of traditional industries is a challenge faced by all countries in the world. In the process of industrial digitalization, China gives full play to the leading and driving role of state-owned enterprises. Through carrying out demonstration projects, China promotes the wide application of new technologies. At the same time, China continuously strengthens the construction of industrial digital infrastructure, builds advanced platforms, and improves the regulatory system. The achievements of China's digital transformation of traditional industries have not only promoted the high-quality development of its own economy, but also contribute Chinese wisdom and Chinese solutions to the digitalization of global industries.

Chapter 6
Digital Society

The digital society is a new type of social formation characterized by the deep integration of digital technologies into social operation and lifestyle. According to the "Overall Layout Plan for the Construction of Digital China" released in 2023, the construction of the digital society includes the universalization of digital public services, the precision of digital social governance, the intelligentization of digital life, and so on. This chapter introduces China's achievements in smart cities, digital villages, smart education, and smart healthcare in recent years.

6.1 Smart City

China stands as the world's most populous nation, with its demographic concentration predominantly within urban centers. The urbanization rate of China embarked on an ascendant trajectory, crossing the 40% threshold in 2003 and surging to 51.27% by 2011. This milestone heralded the advent of an era where urban societies form the primary demographic paradigm. Presently, there exist over 300 cities in China boasting populations exceeding one million, and projections, based on an annual growth rate of 1.2%, foresee that over 70% of the population will live in by the year 2030. The aggregation of dense populations within urban locales poses formidable challenges to essential infrastructure such as urban transportation, municipal services, and power supply. Addressing these challenges necessitates the adoption of crucial measures, with the construction of a smart city through digitalization and informatization emerging as a pivotal solution.

China perceives the concept of smart cities as the advanced phase in the realm of urban informatization. In January 2013, the Ministry of Science and Technology (MOST) and the National Standardization Administration initiated 20 smart city pilot demonstration projects, paralleled by the Ministry of Housing and Urban–Rural Development (MOHURD) launching the first batch of 90 national smart city

W. Liu et al. (eds.), *Digital China*, Introduction to China's S&T Innovation,
https://doi.org/10.1007/978-981-95-6240-4_6

pilot projects. The year 2014 witnessed the issuance of the guideline titled "Guidance on Promoting the Healthy Development of Smart Cities", followed by the introduction of national standards in 2017, including the "Smart City Technical Reference Model" and the "Smart City Evaluation Model and Basic Evaluation Indicators". As of August 2018, all sub-provincial cities and above, along with over 76% of prefecture-level cities and 32% of county-level cities, totaling approximately 500 cities, had explicitly articulated their commitment to or were in the process of constructing innovative smart cities.

6.1.1 Urban Traffic Coordination

Urban transportation stands as a pivotal facet of municipal operational management. Over 400 cities in China have established intelligent traffic command and control centers, amalgamating functions such as emergency response, information gathering, and traffic control. These cities have also implemented oversaw key systems including traffic signal control, traffic guidance, traffic surveillance, and electronic law enforcement. Mega-cities such as Beijing, Shanghai, Guangzhou, and Shenzhen have erected transportation operations coordination centers (TOCC), which can intelligently monitor, coordinate, and command the integrated traffic operations, and achieved comprehensive coordination and management of urban road traffic, rail transit, civil aviation, as well as urban parking and public transportation services.

Significant achievements have been realized through the public transport comprehensive monitoring platform. Taking the city of Shenzhen as an example, since the reform and opening-up, Shenzhen has metamorphosed from a small border town with a permanent resident population of just around 30,000 into an expansive and densely populated city spanning 1997 square kilometers, with a permanent resident population of 17.56 million. It boasts the highest vehicle density in the nation (over 500 vehicles per kilometer) and the third-highest container throughput at its ports nationwide. Shenzhen has pioneered the construction of TOCC in China, encompassing domains of sea, land, air, and rail (see Fig. 6.1). The city's 2700 signal-controlled intersections boast a 98% networking rate, resulting in a 70% reduction in personnel input through optimized timing and a decrease of over 30% in the workland for manual inspections. The deployment of intelligent traffic signal control at over 700 intersections citywide has significantly enhanced traffic management efficiency. In 2020, during the morning rush hour, the average road speed in Shenzhen reached 27.1 km/h, maintaining optimal traffic efficiency among first-tier cities, and the vehicular mortality rate plummeted to 0.62 people per 10,000 vehicles.

Individual-oriented public transport data services become more and more popular. Leveraging extensive data from mobile terminal positioning, smartphone navigation applications can estimate real-time traffic conditions. For instance, Baidu Maps APP integrates data from five major data sources: enhanced location data, real-time satellite orbit data, road data, satellite signals, and mobile sensor data. This integration facilitates minute-level real-time updating of dynamic traffic data for over 380 cities

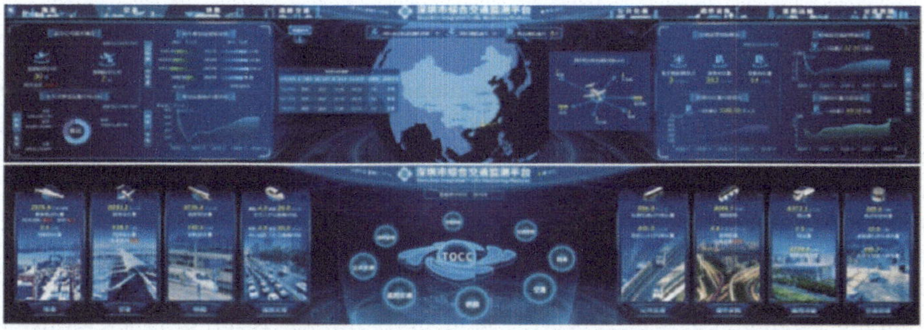

Fig. 6.1 Urban traffic management in Shenzhen: TOCC. *Image Source* Shenzhen Integrated Transportation Operations Command Center

Fig. 6.2 Smart Eye function in the Baidu Maps: urban commuting big data and traffic flow migration trend awareness. *Image Source* Baidu Maps

nationwide, with a location accuracy reaching the lane level (see Fig. 6.2). By the end of 2021, Baidu Maps averaged over 130 billion daily location service requests, covering over 11 million kilometers of roads, and serving more than 600,000 mobile applications.

6.1.2 City Brain

The concept of the City Brain is an innovative initiative originating from China, representing a commendable exploration for urban development worldwide. The

City Brain is a product that combines the architecture of the Internet Brain with the construction of smart cities, constituting a city-level and brain-like complex intelligent mega-system. Spearheaded by the municipal government of Hangzhou, the City Brain was initiated in March 2016, and the City Brain 1.0 was fully operational by October of the same year at the Yunqi Conference in Hangzhou. It can enhance urban operational efficiency and address the intricate challenges inherent in urban management. Presently, there exists a lack of unified understanding of the City Brain among the industrial sector, academia, and various sectors of economic and social development. Some approach it from the perspective of urban transportation, some from the standpoint of urban security, some delve into the realm of urban artificial intelligence hubs, and still, some from the perspective of the brain-like structure, resulting in a multitude of research angles and implementation outcomes.

In November 2020, the Department of Informatization and Industrial Development of the National Information Center proposed that the core of the City Brain was a data-driven artificial intelligence central platform based on ubiquitous sensing, comprehensive interconnection, and digital twinning at the Smart City Expo World Congress. In December 2020, the City Brain Global Standard Research Group, the Research Center on Fictitious Economy & Data Science of the Chinese Academy of Sciences, and the Center for Digital Governance Studies of the China Institute for Innovation and Development Strategy jointly released the "City Brain Global Standard Research Report", outlining nine research directions for global standards related to the City Brain, and research on the "City Brain" continues to deepen. As of 2020, the City Brain had been implemented in over 10 cities, including Hangzhou, Suzhou, and Haikou, and nearly 500 cities nationwide has initiated City Brain construction projects, with an investment exceeding hundreds of billions of yuan.

Hangzhou, as the inaugural city to adopt the City Brain, began its collaboration with Alibaba Cloud in October 2016. The City Brain has been applied to urban transportation, emergency rescue, and public safety (see Fig. 6.3). As of 2017, the City Brain had taken control of 128 signalized intersections in Hangzhou, resulting in a 15.3% reduction in travel time in pilot areas and a saving of 4.6 min on elevated road journeys in the main urban area; in the main urban area of Hangzhou, the City Brain generated over 500 daily event alerts with an accuracy rate of 92%; in Xiaoshan District, the arrival time of ambulances had been halved, with over 120 ambulances reaching the scene in record time.

6.1.3 Digital Twin City

The concept of "Digital Twin" involves mapping a physical entity into virtual space, creating a "Digital Counterpart". In the context of a digital twin city, Internet of Things (IoT) technology, geographic information technology, and intelligent building city models are employed to transform the physical city into a virtual digital city model, enabling functions such as prediction, analysis, and optimization through model simulating operations. Digital twin cities seamlessly integrate the virtual and

Fig. 6.3 City brain in Hangzhou: urban transportation and public safety. *Image Source* rmzxb.com.cn

the real, allowing for the overlap of physical entities and their corresponding twins, facilitating interaction between the digital and physical realms on the foundation of digitalization and informatization.

In June 2021, the MOHURD issued the "Technical Guidelines for the City Information Model (CIM) Foundation Platform" (Revised Edition), reinforcing guidelines for the construction of the "three-dimensional digital foundation" of digital twin cities from a technical implementation perspective. Regions such as Shanghai, Zhejiang Province, and Xiong'an New Area in Hebei Province, expedited the establishment of city-level CIM foundation platforms, exploring and deploying pilot applications for digital twin cities.

Shanghai has initiated the trial operation of a comprehensive city management system based on digital twin technology, achieving a preliminary realization of "an integrated network for urban operations" (see Fig. 6.4). This system integrates access to public security, greening and urban appearance, housing and urban and rural development, transportation, emergency response, ecological environment, health, and other fields, and supports cross-departmental and cross-system joint operations to increase efficiency and empowerment. As of 2022, the Shanghai Urban Operation and Management Center had connected to 220 systems of 72 departments (units) and a number of digital twin application scenarios and continuously summarized and improved them, actively promoted the application of digital twins in infrastructure, historical buildings, communities, cultural tourism, education, healthcare, emergency response, fire protection, and other fields, and created a number of unique digital twin application scenarios in urban safety, urban operation, and innovation leadership. Shanghai has put forward the concept of urban agent, taking a building as the smallest management unit of the city,and has been exploring innovative practices in the transformation of urban digital governance.

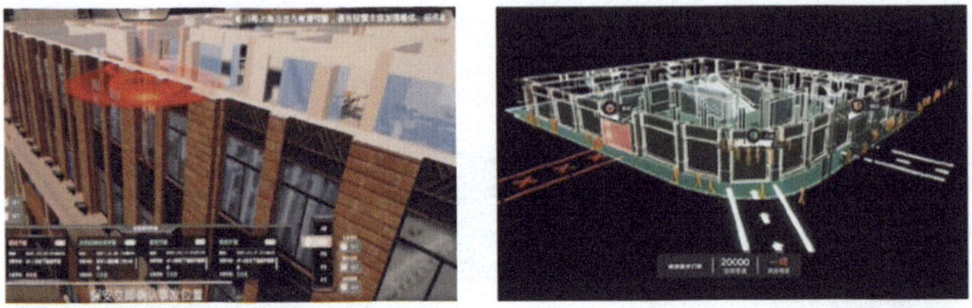

Fig. 6.4 Digital twin city application in Shanghai: digital governance based on urban intelligence. *Image Source* Shanghai Internet Information Office

6.2 Digital Village

From a historical perspective across various nations globally, the development of industrialization and urbanization has often been accompanied by a decline in rural populations and the deterioration of rural economies, leading to social upheaval. Rural areas, typically characterized by vast expanses and sparse populations, face high costs and slow returns in constructing essential infrastructure such as public transportation, information networks, and healthcare. As a socialist country, China places great emphasis on addressing the rural–urban relationship in the process of modernization. Drawing lessons from Western nations experience in rural–urban development, China has formulated the strategic goal of "rural revitalization".

In the 14th Five-Year Plan, China articulated the strategy of "supporting agriculture with industry, promoting rural areas with urban development, fostering a new model of mutually reinforcing relationships between industry and agriculture, and urban and rural areas, characterized by coordinated development and shared prosperity and accelerating the modernization of agriculture and rural areas". China regards the construction of digital villages as an important way to promote rural revitalization and the integrated development of urban and rural areas.

6.2.1 Rural Network Infrastructure

6.2.1.1 Village-To-Village Project

Since the initiation of the reform and opening-up policy, China's ICT industry has made significant strides, yet the development disparities in communications, particularly in revolutionary base areas, ethnic minority areas, national border areas and impoverished areas, have become increasingly prominent. In 2003, nearly 80,000 out of 695,000 administrative villages still lacked telecommunications access. To address the challenges in rural communications development, the Ministry of Information Industry introduced the "Implementation Plan for Universal Service of

Rural Communications—Village Connection Project" (abbreviated as the "Village-to-Village" project) in January 2004, requiring basic telecommunications operators to undertake the construction of communications networks in rural and remote areas through segmented contracting, self-raised funds, and self-operated maintenance. The project was implemented in three phases: the first phase (2004–2005) aimed to achieve "telecommunications access for over 95% of administrative villages", the second phase (2006–2010) aimed to achieve "telecommunications access for every village, and Internet access for every township", and the third phase (2011–2015) aimed to achieve "broadband access for every village" in administrative villages.

Constructing communications networks in remote rural areas presents numerous challenges. On the one hand, these areas are often characterized by remote geographical locations, harsh climatic conditions, and frequent natural disasters, leading to significant difficulties in construction projects and posing extreme challenges to human labor. On the other hand, rural areas, as a whole, exhibit relatively backward economic development and low population density, giving rise to high construction and ongoing operational costs for both optical fiber networks and mobile communications base stations. Furthermore, the number of users is limited, resulting in low market returns. According to estimates, the cost of connecting each administrative village by optical fiber networks in the "three areas and three prefectures"[1] is on average four times that of the eastern and central regions of the country; the construction cost of each base station is, on average, 3.3 times that of the eastern and central regions of the country; in some remote areas, this difference reaches approximately tenfold.

As the most populous developing country globally, China leverages its institutional advantages of concentrating efforts to address major issues. The state-owned telecom operators employ the "segmented contracting" method to advance the "Village-to-Village" project, allocating the funds needed for universal telecom services among the six telecom operators at that time, including China Telecom, China Netcom, China Mobile, China Unicom, China Satcom, and China Tietong. The allocation proportions for each operator are consistent with their respective shares of the total revenue and profits among all telecom operators, with revenue and profit weights each accounting for 50%.

After over a decade of implementation, the phased goals of the "Village-to-Village" project have been essentially achieved. The construction and service levels of rural communications have been significantly improved, with the scope of universal communications services gradually expanding from voice services to Internet services. By the end of 2014, the annual tasks of the "Village-to-Village"

[1] The "three areas" refer to the Xizang Autonomous Region, the Hotan, Aksu, Kashgar prefectures and the Kizilsu Kirgiz Autonomous Prefecture in southern Xinjiang, and prefectures and counties with large Tibetan populations in the provinces of Sichuan, Yunnan, Gansu and Qinghai. The "three prefectures" refer to the Liangshan Yi Autonomous Prefecture in Sichuan Province, the Nujiang Lisu Autonomous Prefecture in Yunnan Province, and the Linxia Hui Autonomous Prefecture in Gansu Province.These areas have harsh geographical conditions and lay behind in economic development,and are designated as national-level improverished regions.

project had been exceeded, achieving new progress in the popularization of telecommunications services. Specifically, the proportion of administrative villages with broadband access nationwide reached 93.5%, the proportion of villages with more than 20 households having telecommunications access reached 95.8%, over 3000 townships and villages and 150,000 administrative villages completed broadband construction, and 18,000 impoverished villages achieved Internet coverage, along with 6400 rural schools and public institutions gaining broadband access.

6.2.1.2 Internet-Based Poverty Alleviation Action Plan

In 2012, China initiated the poverty alleviation project, aiming to address the poverty issues affecting approximately 122 million rural residents nationwide. Following the "Village-to-Village" project, an unprecedented nationwide campaign for Internet-based poverty alleviation was launched across the country. Its goal is to bring advanced optical fiber broadband and 4G networks to impoverished villages, achieving parity in Internet access and speed between rural and urban areas.

On October 14, 2015, the State Council executive meeting decided to increase central financial investment, guide localities to strengthen policy and financial support, encourage telecom operators, broadcasting enterprises, and private capital to participate fairly in rural broadband construction and operation and maintenance through competitive bidding and explore market-oriented approaches such as public–private-partnership (PPP) and entrusted operation to mobilize the participation of various entities.

In 2016, the Office of the Central Cyberspace Affairs Commission, the National Development and Reform Commission (NDRC), and the State Council Leading Group Office of Poverty Alleviation and Development jointly issued the "Action Plan for Internet-based Poverty Alleviation", which proposed five major projects: network coverage project, rural e-commerce project, network-based intelligence support project, information service project, and network public welfare project. Among them, the network coverage project, that is, the pilot project of universal telecommunications services, mainly involves the most challenging construction tasks left by the "Village-to-Village" project, and are mainly distributed in remote areas with high mountains and valleys and few people (see Fig. 6.5).

The state-owned telecom operators and other key industries spared no effort, with millions working day and night, creating a world miracle in rural communications coverage. As of 2020, China had implemented six batches of universal telecom service pilot projects, covering 27 provinces, autonomous regions, and municipalities nationwide, and completed the construction of optical fiber networks for 130,000 administrative villages and 4G base stations in 50,000 rural areas, including optical fiber networks in 43,000 impoverished villages and 4G base stations in 15,000 impoverished villages. The national coverage ratio of optical fiber networks and 4G networks in administrative villages exceeded 98%, and the average download speed of rural optical fiber broadband surpassed 100 Mbps, achieving parity in Internet access and speed between rural and urban areas. The goal of "broadband network

Fig. 6.5 Internet-based poverty alleviation project: harsh conditions for laying out communications networks. *Image Source* NI GN, *China's Road to Science and Technology Information Volume Intelligent Connection of Everything*, 2021

coverage for over 90% of impoverished villages" proposed in the "13th Five-Year Plan for Poverty Alleviation" was achieved ahead of schedule and was surpassed. Information superhighways comparable to those in urban areas have been established in the once-isolated deep mountains, allowing more and more people in remote urban areas to share the digital dividends of the information era and embark on the path to prosperity and well-being.

In the Internet-based poverty alleviation action plan, China's state-owned telecom operators have made substantial investments in rural areas. As of 2019, China Telecom had invested more than 100 billion yuan in rural areas, achieving 100% 4G network coverage in townships nationwide, broadband access in more than 91% of administrative villages, and 90% optical fiber broadband coverage in townships and towns nationwide; China Mobile had invested more than 80 billion yuan, providing telephone access in 122,000 villages and broadband access in 840,000 administrative villages, with a 4G coverage rate of more than 98% in administrative villages nationwide; China Unicom had achieved 100% network coverage in more than 90,000 townships amd towns nationwide, and the total number of administrative villages covered had exceeded 460,000.

China Telecom, China Mobile, and China Unicom have launched large-scale initiatives to offer discounts, ensuring that residents in remote areas not only have access to broadband networks, but can also afford them and utilize them effectively.

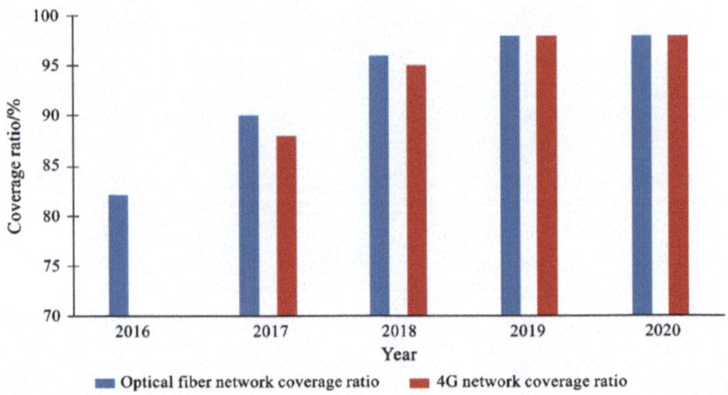

Fig. 6.6 Optical fiber network and 4G network coverage in China's administrative villages (2016–2020). *Data Source* Ministry of Industry and Information Technology of the People's Republic of China

In 2020, China Telecom had launched poverty alleviation packages at 35,000 business outlets across the country, offering discounts totaling more than 1.9 billion yuan since 2019, built a big data management platform for targeted poverty alleviation, and served 39 million impoverished residents in 17 provinces, autonomous regions, and municipalities; China Mobile had launched heavily discounted exclusive "poverty alleviation packages" and mobile phone purchase subsidies for impoverished residents, benefiting more than 14 million registered poor households and offering profits of 3 billion yuan, and donated its own-brand mobile phones and other terminal devices to residents in poverty-stricken areas; China Unicom provided tariff discounts for poverty-stricken areas, launched more than 150 exclusive discount packages, and cumulatively reduced and exempted communications fees by 367 million yuan, helping more than 400,000 poor households to successfully get rid of poverty. The optical fiber network and 4G network coverage in China's administrative villages from 2016 to 2020 is shown in Fig. 6.6.

Through a series of rural network infrastructure developments, China has progressively narrowed the digital divide between rural and urban areas. In 2021, broadband coverage was achieved in every administrative village, with a nationwide broadband coverage rate of 100%. Optical fiber network and 4G network coverage rate both exceeded 99%, essentially realizing the goal of "same network access and speed in rural and urban areas". The extension of 5G to rural areas has been expedited. As of 2022, China had established and activated a total of 1.968 million 5G base stations, and the 5G network had covered all prefecture-level city urban areas, county-level city urban areas, and 96% of township areas, achieving "5G coverage in every county". As of 2021, rural residents, on average, had 229 Internet-enabled mobile phones per 100 households, marking a 4.4% increase from the previous year. As of June 2022, the Internet users in rural areas had reached 293 million, with a rural Internet penetration rate of 58.8%, narrowing the urban–rural Internet penetration gap by nearly 15 percentage points compared to the 13th Five-Year Plan period.

6.2.2 Rural E-commerce

The Chinese government considers the development of rural e-commerce as a primary means to advance the digital economy in rural areas. Governments of various levels have initiated and organized diverse activities to promote rural e-commerce, such as the "Internet Plus" project for the online sale of agricultural products, the "Digital Business Empowers Agriculture" project, and the "Great Country Craftsman" National Farmers Skills Competition. In recent years, the Ministry of Commerce (MOFCOM) and other departments have continued to promote rural e-commerce, achieving remarkable results. As of 2023, more than 20 billion yuan of central financial funds had been arranged to support a total of 1489 counties, and construct more than 2600 county-level public service centers for e-commerce and logistics distribution centers and more than 150,000 rural e-commerce and express delivery service sites. Express delivery services have continued to extend to primary rural areas. In 2021, more than 80% of villages were covered by the "Express Delivery to Villages" initiative, with the total volume of express parcels in rural areas reaching 37 billion.

By the end of 2021, 36.3% of key agricultural leading enterprises at or above the municipal level had engaged in sales through e-commerce. Agricultural processing enterprises utilizing e-commerce reported a 10.8% increase in operating revenue compared with the previous year. E-commerce plays a crucial role in facilitating the sales of agricultural products in poverty-stricken areas, contributing significantly to preventing a return to poverty on a large scale. In 2022, China's rural online retail volume reached 2.17 trillion yuan, marking a 3.6% increase from the previous year. China's rural online retail value from 2016 to 2022 is shown in Fig. 6.7.

Live streaming for product promotion, known as "live commerce", has become a highlight in rural e-commerce. Under the coordination of relevant departments such as the Regional Revitalization Department of the NDRC and the Public

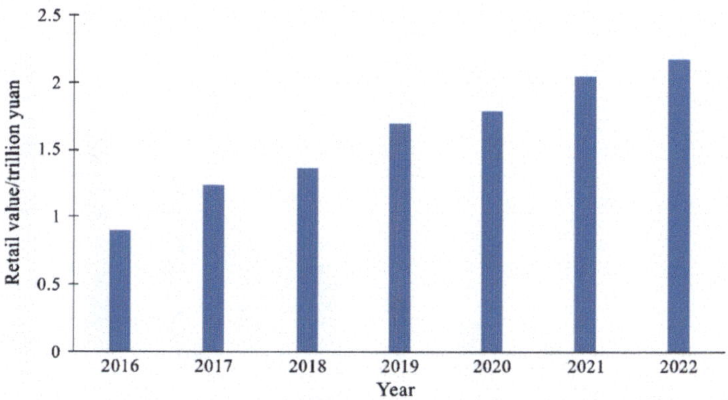

Fig. 6.7 China's rural online retail value (2016–2022). *Data Source* Ministry of Commerce of the People's Republic of China

Service Department of the National Radio and Television Administration, Alibaba Group's Taobao platform launched the "Village Live Streaming Program" in 2019, providing training for farmers in skills such as online live streaming and online traffic management. As of September 2021, the program had incubated over 110,000 farmer livestreamers, conducted more than 2.3 million live streaming sessions to assist farmers, covered over 2000 counties in 31 provinces, autonomous regions, and municipalities, and driven agricultural product sales exceeding 5 billion yuan, creating employment opportunities for over 200,000 people.

The "Taobao village" develops rapidly. It is a term for rural e-commerce and regional agglomeration. There are three statistical standards for Taobao villages: (1) Business location. In rural areas, the administrative village is used as the unit. (2) Sales scale. The annual e-commerce sales of the whole village reaches 10 million yuan. (3) Online merchant scale. The number of active online stores in the village reaches 100, or the proportion of active online stores reaches 10% of the local households. Since the emergence of the first three Taobao villages in 2009, Taobao villages have achieved exponential growth. As of 2022, Taobao villages had covered 28 provinces, autonomous regions, and municipalities, and 180 cities (regions) across the country, with a total number of 7780 (see Fig. 6.8). The emergence of Taobao villages and their increase in number represent the continuous improvement of the scale of grass-roots e-commerce operations based on administrative villages, attracting more and more talents to take root in rural areas and promote the prosperity of rural industries.

Rural e-commerce represented by Taobao villages has become one of the important ways for the Chinese government to eliminate poverty in rural areas. Taking Heze City, Shandong Province as an example, it promotes the "e-commerce poverty alleviation" model, deepens cooperation with e-commerce platforms such as Alibaba and JD.com, and carries out targeted training such as Taobao store operation skills improvement, new media operation, and live streaming. As of November 2018,

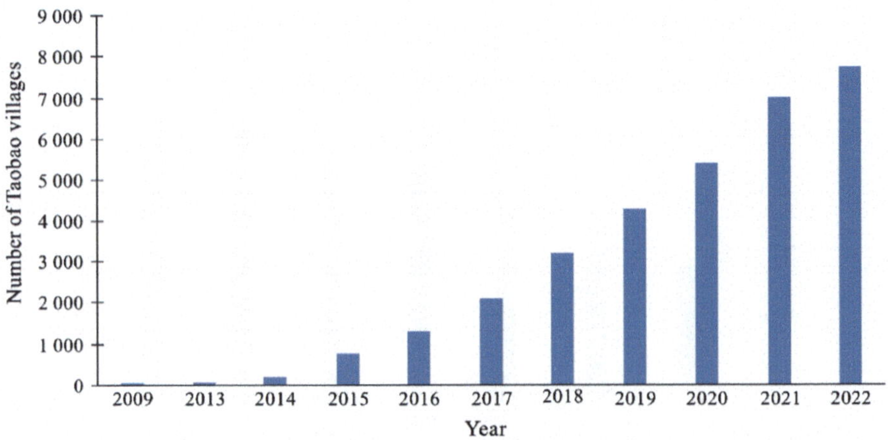

Fig. 6.8 Number of Taobao villages in China (2009–2022). *Data source* Ministry of Commerce of the People's Republic of China

Heze City had developed e-commerce in 563 poor villages, benefiting more than 25,000 impoverished residents, of which 43 poor villages have developed into Taobao villages and achieved poverty alleviation for the entire village. In October 2019, the Heze Poverty Alleviation and Development Office won the 2019 National Poverty Alleviation Award for Organization Innovation. In November 2019, the World Bank, Alibaba Group, and the China International Development Knowledge Center released the "E-commerce Development: Experience from China". The report affirmed that the pioneering spirit of online merchants is the most basic and fundamental driving force for the development of Taobao villages. It expected that the experience of Taobao villages will be more widely disseminated and applied in the central and western regions, so as to change the situation of more backward villages and towns, and it also expected that the experience of Taobao villages can be replicated and spread in the vast developing countries.

Internet-based informatization has helped accelerate the development of rural tourism, leisure agriculture, and homestay economy. As of September 2022, the Ministry of Agriculture and Rural Affairs (MARA) had released more than 70 promotions of rural leisure tourism boutique scenic spots and routes through its official website, covering 211 rural leisure tourism routes in 148 counties (cities, districts) of 31 provinces, autonomous regions, and municipalities across the country, and utilized the "Want to Go to the Countryside" mini-program to promote 681 rural leisure tourism boutique routes, covering over 2500 high-quality scenic spots and other high-quality resources. The number of people returning to their hometowns to start businesses is growing rapidly. In 2021, that number reached 11.2 million in China, an increase of 10.9% over 2020.

6.2.3 Rural Public Services

The level of comprehensive rural grass-roots governance has been continuously improved. The "Internet + Grass-roots Social Governance" action has been implemented in depth, and various regions have actively promoted the construction and open sharing of grass-roots social governance data resources. By the end of 2021, China had 483,000 village-level comprehensive service stations that use information technology to carry out basic public services and public utility services, with a coverage rate of 86.0% of administrative villages. In 2021, the administrative village coverage of the Public Safety Video Application System reached 80.4%. Notably, critical water level alarm monitoring and panoramic surveillance were installed in rural water areas, effectively preventing water-related accidents. By leveraging the Child Welfare Management Information System, a comprehensive understanding of the care service recipients in rural areas has been obtained. From July 2021 to June 2022, information on 755,000 left-behind children was collected, further enhancing the precision of child welfare and protection of minors in rural areas.

Information services for agriculture, rural areas, and rural residents have become more accessible and profound. By the end of 2021, 467,000 beneficial agricultural information centers had been established and operated across the nation, collectively providing various information services to 980 million individuals. Agricultural technology services have transitioned from the fields to online platforms, with innovative services methods such as the 12,316 htline and the National Agricultural Science and Technology Service Cloud Platform Continuously evolving to serve farmers. As of August 2022, the registered users on the National Agricultural Science and Technology Service Cloud Platform had surpassed 13 million, accumulating over 3.5 billion visits, with a daily service engagement exceeding 4 million individuals, and the online question-answering response rate maintained over 92%. Notably, in 2021, the nationwide number of new agricultural business entities (including farmer cooperatives and family farms) receiving information technology-based agricultural technology promotion services reached 2.233 million.

The rural online cultural life is vibrant and diverse. The Internet has become an important channel for everyone to participate in and experience the Chinese farmers harvest festival. In 2022, China Central Radio and Television Station created the first immersive online harvest festival gala "2022 Online Harvest Evening". "Internet Plus" mass cultural activities are booming. During the New Year's Day and Spring Festival in 2022, the National Public Culture Cloud Platform launched an online "Village Evening" special, live broadcasting 127 selected events from various places, with 148 million online participants. The "June 6th" Miao Chixin Festival basketball game spontaneously organized by the villagers of Taipan Village, Taijiang County, Guizhou Province has become a viral sensation, affectionately called Village Basketball Association (VBA) by netizens. The related online broadcasts and short videos garnered over 100 million online viewers (see Figs. 6.9).

The continuous advancement of digital technology has facilitated the ongoing exploration and realization of the cultural, social, and economic values inherent in agricultural civilization. The digitalization of intangible cultural heritage resources within China's traditional villages is steadily progressing, incorporating rural cultural forms with significant value and distinctive characteristics into the overall protection scope of national pilot zones for ecological conservation. In 2021 and 2022,

Fig. 6.9 Live streaming of VBA in Taipan Village, Taijiang County, Guizhou Province. *Image Source* Tianyan News, Xinhua News Agency, CCTV News, CCTV Video

significant support was directed towards the digital preservation efforts of intangible cultural heritage resources in 364 traditional Chinese villages. As of June 2022, the digital museums for traditional Chinese villages had meticulously gathered and organized information on 6819 such villages. The construction of 658 individual village exhibits had been completed, establishing a comprehensive database of traditional villages that encompasses various data types, including panoramic roaming, texts and images, audio-visual elements, and real-life models.

6.3 Smart Education

China has made great achievements in national education. Firstly, the level of education popularization has achieved a historic leap. By the end of 2021, there were 529,300 schools of all levels and types in China, with 291 million students enrolled. There were 207,000 schools in the compulsory education stage,[2] with 158 million students enrolled, and the comprehensive popularization of compulsory education had been achieved; there were 22,000 high schools with 39.17 million students and a gross enrollment rate of 91.4%; there were 3012 higher education schools with a total enrollment of 44.3 million people and a gross enrollment rate of 57.8%. Secondly, new achievements have been made in educational equity. As of 2022, 2895 counties in China had achieved basic balance in compulsory education and established a student aid policy system covering "all educational stages, all schools, and all economically disadvantaged students" to ensure that "no student loses educational opportunities due to family economic difficulties".

Education informatization helps promote the sharing of high-quality educational resources, promote educational equity, and improve educational quality. The Ministry of Education (MOE) attaches great importance to education informatization and has carried out long-term and continuous construction of education informatization. In 2012, the MOE issued the "Ten Year Development Plan for Education Informatization (2011–2020)", setting the goal of achieving "Three Universal Accesses, and Two Internet-based Platforms", which includes "broadband network access for all schools, high-quality resources for all classes, and online learning space for everyone", as well as the educational resource public service platform and the educational management public service platform. In 2018, the MOE released the "Education Informatization 2.0 Action Plan", which proposed the development goal of "Three Coverages, Two Enhancements, and One Major Platform". "Three Coverages" means full coverage of teaching applications for all teachers, learning applications for all school-age students, and digital campus development for all schools. "Two Enhancements" means enhancing the application level of ICT in education and digital literacy

---

[2] Since the promulgation of the "Compulsory Education Law of the People's Republic of China" in 1986, China has established a compulsory education system that spans nine years (primary and secondary education stages), all school-age children and adolescents to receive compulsory education.

among teachers and students. "One Major Platform" is to build a large-scale "Internet + Education" integrated platform.

After years of education informatization construction, China's education informatization development stage has shifted from application popularization to integrated innovation. Utilizing information technology to promote the digital transformation of Chinese education and achieve high-quality development has become a development task in the new era. In February 2019, China released the "China's Education Modernization 2035", which was the first medium-and long-term strategic plan with the theme of education modernization. It proposed to "accelerate the transformation of education in the information age", with a focus on the construction of intelligent campuses and the use of modern technology to accelerate the reform of talent training models. In July 2021, the MOE and five other ministries issued the "Guidance on Promoting the Construction of New Education Infrastructure and Building a High-quality Education Support System", proposing a new education infrastructure system including information networks, platform systems, digital resources, smart campuses, innovative applications, and trustworthy security.

6.3.1 Digital Campus Facility

The network conditions of schools at all levels and types have improved significantly. In 2012, the MOE proposed the construction goal of "broadband network access for all schools". As of 2022, the Internet access rate of primary and secondary schools (including teaching points[3]) in China had reached 100% (see Fig. 6.10), an increase of 75 percentage points from 2012; more than 75% of schools were covered by wireless networks, 99.5% of schools had multimedia classrooms, with a total number of more than 4 million, and the number of teacher and student terminals equipped in schools had exceeded 28 million; the number of digital terminals per 100 students in China's primary and junior high schools was 14.9 and 21.0 respectively, an increase of 8.4 and 10.6, respectively, compared with ten years ago; the digital teaching conditions in primary and secondary schools had been comprehensively upgraded, and a ubiquitous learning environment with complete network coverage and the integration of offline multimedia teaching space and online teaching space had basically been formed.

With the improvement of campus network conditions, high-quality educational information resources can be effectively shared. Taking the Ningxia Hui Autonomous Region, which is located in the hinterland of northwest China as an example, the issues of unbalanced and insufficient education development there is relatively prominent. In 2018, Ningxia was approved to build a national "Internet + Education" demonstration zone. The whole region achieved 200 M network broadband access in schools and built 29,000 digital classrooms. With the help of high-speed and

[3] Teaching points are small-scale, incomplete primary schools established in remote and educationally under-resourced areas, primarily utilizing multi-grade classes (combined grades in one classroom). They typically cover only lower grades, with fewer than 20 students per point.

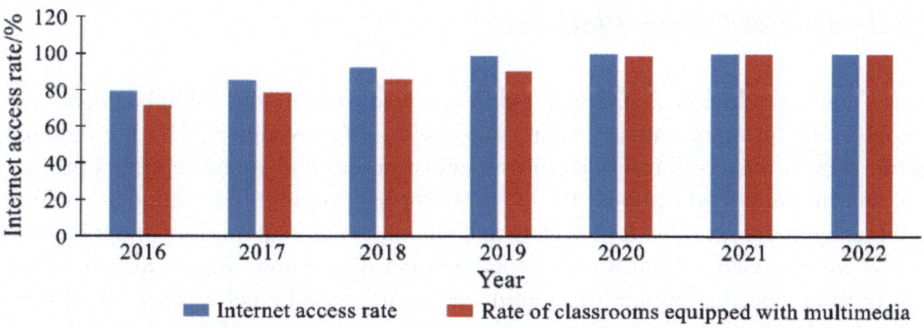

Fig. 6.10 Campus network construction of China's primary and secondary schools (2016–2022). *Data Source* Ministry of Education of the People's Republic of China

complete network infrastructure, more than 500 schools in Ningxia had conducted online pairing-up[4] with high-quality schools in Beijing, Fujian, and other places, and all rural primary and secondary schools cooperated with urban schools online, effectively alleviating the problem of teachers shortages in under-resourced schools through cross-school collaborative teaching and online teaching assistance. The index of education informatization development in Ningxia rose from 15th in the whole country in 2017 to 6th in 2020. By 2023, the number of Ningxia users registered at the Smart Education of China platform had reached 1.25 million, covering 100% of teachers and 64.9% of students in primary and secondary schools in the region.

At present, China has established a complete digital campus standard system, covering the digital construction standards of different types of schools such as primary and secondary schools, vocational schools, and colleges and universities. The construction aspects of digital campus covers online education, education management, education evaluation, application services, cybersecurity and other aspects. In 2012, the Ministry of Education launched activities such as "100 Digital Campus Demonstration Schools" evaluation to further implement the standards; provincial education authorities also regularly carry out evaluation and assessment work on digital campus construction. The current focus of campus informatization construction has shifted from "digital campus" to "smart campus". For example, in 2023, the Beijing Municipal Education Commission issued the "Beijing Primary and Secondary School Smart Campus Construction Specifications (Trial)", proposing that the content of smart campus construction includes intelligent environment, application integration and innovation, school education data and applications, Internet services and applications, digital literacy and skills, security and operation services, etc., with a focus on co-construction and sharing and smart applications.

[4] Online pairing-up serves as a specific implementation of pairing assistance. China has established this mechanism between developed and underdeveloped provinces to deliver targeted support through cross-regional collaboration in funding, technology, and resource allocation, effectively narrowing regional development gaps. This approach leverages the institutional strengths of socialism with Chinese characteristics, achieving coordinated development across regions via government-led resource redistribution.

6.3.2 Digital Course Platform

China has established a national public service system for digital educational resources for primary and secondary schools. As of November 2020, the national public service system for digital educational resources had connected to 84 provincial and municipal educational resource platforms, providing educational teaching and management services for teachers and students; the total number of application visits had exceeded 300 million, the total number of resource sharing had exceeded 320 million, and the number of monthly active users had reached over 60 million; more than 100 million teachers and students had created their online learning spaces, and nearly half of teachers used online learning spaces for teaching and research; to supply digital educational resources for rural schools and teaching points, a special project had been implemented to share 6948 h of digital resources in subjects such as English, music, and art, as well as a total of 50 million resources aligned with all subject textbooks in the basic education stage.

China has established a series of online course platforms for higher education. In 2013, Chinese universities began to build massive open online course (MOOC). Different kinds of online teaching modes, such as cross-school and cross-regional online teaching, "1 (MOOC) + M (universities) + N (students)"collaborative teaching and online and offline hybrid teaching, "MOOC + SPOC[5] + flipped classroom" have been realized in the university classrooms. In 2019, the Higher Education Department of the MOE issued a call for "MOOCs for the West", and a total of 172,900 MOOCs and customized course services have been provided to the under-resourced universities in the western China, helping the western region to carry out 3.27 million blended teaching courses, with 376 million students participating in learning and 1.71 million teachers in the western region participating in application training.

In early 2020, the sudden outbreak of the COVID-19 pandemic triggered an unprecedented large-scale online teaching practice. The MOE launched the campaign of "suspension of classes but not learning", opening up a National Primary and Secondary School Network Cloud Platform to ensure home-based learning for students in areas without network or with slow network speeds through Classroom in the Air (TV version). Relying on the existing educational platforms and resources, teachers at all levels and types of schools actively carried out teaching activities (see Fig. 6.11). In February 2020, the MOE's National Primary and Secondary School Network Cloud Platform was successfully launched, coordinating 7000 servers and 90 T of bandwidth, capable of supporting simultaneous online access for 50 million students. As of May 2020, the National Primary and Secondary School network Cloud Platform had accumulated 2.07 billion views and 1.71 billion visits. Additionally, 27 provinces, autonomous regions, and municipalities opened provincial online learning platforms and guided certain cities, counties and schools with necessary infrastructure to make effective use of their local educational resources.

---

[5] Abbreviation for small private online course, a hybrid teaching model that combines physical classroom instruction with online learning.

Fig. 6.11 During the epidemic, students learned online courses in various ways. *Image Source* Wuhan Radio and Television Station

During the epidemic, 1454 colleges and universities in China carried out online teaching, 1.03 million teachers taught 1.07 million courses online, and a total of 17.75 million students studied online. Supported by the Secretariat of the World MOOC and Online Education Alliance, Chinese universities have organized more than 10 global online education dialogues, opened 168 global integrated courses, implemented mutual recognition of credits with 13 world-renowned universities in 11 countries, and launched 8 English global open courses, attracting 7.3 million learners worldwide, with international online education cooperation and exchanges continuing to deepen. On March 13, 2020, United Nations Educational, Scientific and Cultural Organization (UNESCO) released a list of remote learning solutions to the world, recommending 27 learning applications and platforms that can be obtained freely worldwide, among which iCourse and Alibaba DingTalk from China were selected (see Fig. 6.12).

In recent years, China has further integrated online digital education resources and platforms such as the National Digital Education Resource Public Service System and the National Primary and Secondary School Network Cloud Platform. In March 2022, the National Smart Education Public Service Platform for school students was officially launched, covering three major resource platforms: primary and secondary schools education, vocational education, and higher education, and providing 26 government services. The platform has brought together 44,000 basic education course resources, 6628 high-quality online vocational education courses, and 27,000 quality courses in higher education. At the same time, public online course platforms opened to the public are also booming. As of November 2022, there were more than 61,900 MOOCs available online, with 402 million registered users and 979 million learners. China ranked first in the world in terms of the number of MOOCs and the number of learners, and maintains a rapid growth trend. Some MOOC platforms have introduced international versions targeting overseas learners, providing more than 900 multilingual course resources and necessary teaching services to learners from all over the world freely. Nearly a thousand online courses in 14 languages has been made available to college students and learners around the world, with a total of 670,000 learners worldwide.

Fig. 6.12 China's online course platform was included in the UNESCO recommended remote learning solutions. *Image Source* UNESCO

6.3.3 *Internet Education Application*

The "three classrooms" are three types of online collaborative teaching proposed by the MOE for primary and secondary schools in educationally disadvantaged areas, including "special delivery classroom", "famous teacher classroom" and "online classroom from famous school" (see Table 6.1).

In 2012, the MOE first proposed to develop "three classrooms" at the "Education Informatization Pilot Work Symposium", aiming to improve the quality of education in remote areas and promote the sharing of high-quality teaching resources. In 2014, the MOE pointed out in the "Implementation Plan for Building an Effective Mechanism to Expand the Coverage of High-quality Education Resources by Informatization" that it was necessary to promote educational equity and improve the quality of education through various forms such as "three classrooms". In 2016, the "13th Five-Year Plan for Education Informatization" proposed actively promoting the construction of "special delivery classroom" to consolidate and deepen the "full coverage of digital teaching resources at teaching points". In March 2020, the MOE issued the "Guidance on Strengthening the Application of Three Classrooms", proposing the fully realization of "three classrooms" by 2022. In April 2022, the National Center for Educational Technology released innovative cases and shortlisted cases of the application of "three classrooms", promoted typical experiences and practices, and played a demonstration and leading role. The sample application of "three classrooms" is shown in Fig. 6.13, 6.14 and 6.15.

Table 6.1 Three types of online collaborative teaching for educationally disadvantaged areas

Applications	Objectives	Target users	Construction contents	Required facilities
Special delivery classroom	Solve the problem of teacher shortage in rural discouraged schools and teaching points, and the inability to offer the national prescribed courses	Students in rural discouraged schools and teaching points	Use online courses or synchronous classes to help them complete the national curriculum	Live streaming and recording system, remote multimedia display equipment, etc
Famous teacher classroom	Assist skill development for teachers in rural areas	Teachers in rural discouraged schools	Establish an online training community to improve the skills of ordinary teachers by leveraging the expertise of excellent teachers	Interactive recording and broadcasting system, online teacher training platforms, etc
Online classroom from famous school	Narrow the disparity in education quality between regions, urban and rural areas, and schools	Teachers and students in rural discouraged schools	Taking high-quality schools as the main body, promote the sharing of high-quality educational resources in the region through online schools, online courses, etc	Smart classrooms, cross-school teaching platforms, etc

6.3.4 Smart Education Application

The era of artificial intelligence (AI) has brought all-round challenges to education. China attaches great importance to the impact of AI technology on education. In July 2017, the State Council issued the "New-Generation AI Development Plan", which clearly proposed the development of smart education. In April 2018, the MOE issued the "Innovation Action Plan for AI in Higher Education Institutions" and the "Action Plan for Education Informatization 2.0", which further clarified the

Fig. 6.13 Application of the "special delivery classroom": students at the teaching point in Ganzhou Economic Development Zone, Jiangxi Province interacted with students at the teaching end. *Image Source* China Education News

Fig. 6.14 Application of the "famous teacher classroom": Xinhu Primary School in Sanmen County, Zhejiang Province carried out teaching and research activities. *Image Source* China Education News

integrated development of AI and education, and launched a pilot project to promote the construction of the teaching team with AI.

AI education is gradually entering school classrooms. The compulsory module of the "General High School Information Technology Curriculum Standards (2017 Edition, Revised in 2020)" compiled by the MOE has added basic knowledge of AI. Primary and secondary schools in various places are actively exploring the gradual popularization of information technology (IT) and AI technologies to lower-grade students. Schools have carried out the maker education and STEAM (science, technology, engineering, art and mathematics) education based on the facilities of intelligent robots. Some schools introduced the intelligent robot programming courses through IT courses and school-based elective courses established extracurricular interest groups, and offered AI popularization courses in some specified grades,

Fig. 6.15 Application of the "online classroom from famous school": exchange and training in Three Schools Alliance in Jiangxi Province and Zhejiang Province. *Image Source* China Education News

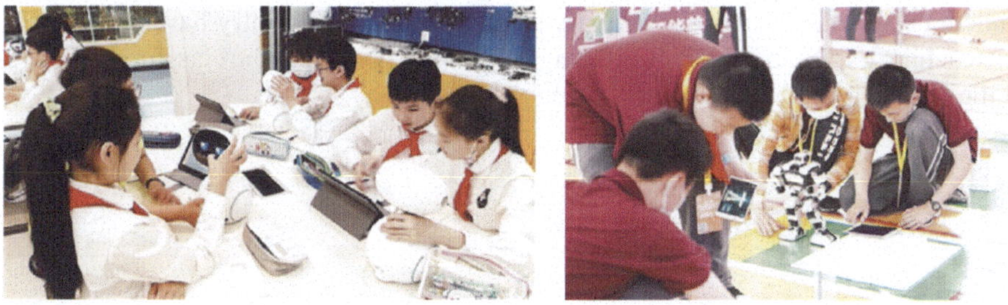

Fig. 6.16 AI course teaching and competition activities in secondary schools. *Image Source* China Education News

bringing AI courses into the classrooms of primary and secondary school students (see Fig. 6.16).

AI technology has also been applied to assist teaching in the school education. In 2016, iFLYTEK and the MOE Examination Center (now the National Education Examination Authority) jointly established a joint laboratory to explore and develop intelligent marking technology for oral English and composition in the college entrance examination. In the first half of 2017, intelligent marking technology was applied to the written examinations of the College English Test Band 4 and Band 6, as well as the high school entrance examination, college entrance examination, and academic examination in 25 regions, covering nearly 7 million candidates. It can detect blank test papers, suspected duplicate test papers, and other problematic test papers, reducing the human labor workload by 20%. According to the "2022 Artificial Intelligence Education Blue Book" jointly released by East China Normal University, China Academy of Educational Sciences, Tencent Research Institute, and Tencent Education, among the 28,000 teacher samples in 25 provinces, autonomous

Fig. 6.17 Primary schools introduced humanoid robots to assist classroom teaching. *Image Source* China Education News (left), People's Daily (right)

regions, and municipalities across the country, 58% of teachers have used AI technology in the subjects they teach, 72% of teachers have adopted computer vision technology for graph recognition, and more than half of teachers have used related tools such as intelligent homework assignments, smart classrooms, and intelligent student performance analysis. Primary schools across the country have also carried out teaching explorations by introducing humanoid robots into the classroom to assist in the teaching process, as shown in Fig. 6.17.

6.4 Smart Healthcare

China has made great achievements in national health. In 2019, the average life expectancy of Chinese residents reached 77.3 years, and the maternal mortality rate, infant mortality rate, and mortality rate of children under 5 dropped to 17.8/100000, 5.6‰, and 7.8‰, respectively. The main health indicators were generally better than the average level of middle-and high-income countries. The framework of the basic medical and health system with Chinese characteristics has been established, and the accessibility of medical and health services has been continuously improving. As of 2020, the number of beds in medical and health institutions per thousand people had reached 6.46, 84% of county-level hospitals had reached the level of second grade hospitals and above,[6] and nearly 90% of families were able to reach the nearest medical point within 15 min. The level of equalization of basic public health services in China has been further improved, with 14 major categories of national basic public health services provided free of charge to all urban and rural residents.

[6] According to the standards of the National Health Commission, the hospitals in China are classified into 3 grades and 10 levels, based on their scale, facility conditions, service scope, etc. In general, first-grade hospitals provide health services to local community residents; second-grade hospitals are regional hospitals that serve multiple communities; third-grade hospitals are cross-regional medical centers that can provide services to patients from local cities, provinces, or even across country.

With the development of China's economy and society, the focus of China's health career in the new era has shifted from disease treatment to universal health. In October 2016, the CPC Central Committee and the State Council promulgated the "Outline of the Healthy China 2030 Plan", which established the new theme of shifting from "focusing on treating diseases" to "focusing on People's health". However, China faced the basic national conditions of uneven distribution of medical resources, large differences in service levels, and a continuously widening gap between supply and demand of medical and health services. The 2020 China Health Statistics Yearbook showed that the number of hospitals, third grade hospitals, health technicians, and practicing physicians in the eastern region was significantly higher than that in the central and western regions. Promoting the informatization and intelligent upgrading of medical and health services is an important way to get rid of the dilemma of uneven distribution of medical resources. In November 2022, the National Health Commission (NHC), the National Administration of Traditional Chinese Medicine, and National Disease Control and Prevention Administration jointly promulgated the "14th Five-Year Plan for Universal Health Informatization", proposing that in the face of the opportunities and challenges brought about by digital transformation, it is necessary to further consolidate the new infrastructure of universal health informatization, cultivate new forms of health services, and enhance new momentum for the development of the health industry.

6.4.1 Digital Healthcare Infrastructure

China's hospital informatization has developed through several stages. It entered the initial stage in the 1990s. The focus of informatization during this period was on management informatization, computerizing traditional business management models (including hospital financial management, fee management systems, etc.) and realizing the application of computer technology in the medical and health system. In 2010, it entered the in-depth application stage. The focus then was to realize various clinical information systems with electronic medical record system as the core, including clinical examination information management system, medical imaging information management system, etc. (see Fig. 6.18). After 2015, it entered the stage of hospital integration platformization, that is, to realize information exchange within the hospitals and promote the in-depth application of electronic medical record.

China has formulated a series of standards for the informatization construction of medical and health institutions across the country. From 2018 to 2019, the NHC successively issued the "Standards and Specifications for National Hospital Informatization Construction (Trial)", "Management Measures for the Grading Evaluation of the Application Level of Electronic Medical Record Systems (Trial)", "Grading Evaluation Standards for the Application Level of Electronic Medical Record Systems (Trial)", and "Standards and Specifications for National Basic Medical and Health Institutions Informatization Construction (Trial)", which clarified the construction content of informatization service standards for various business processes of primary

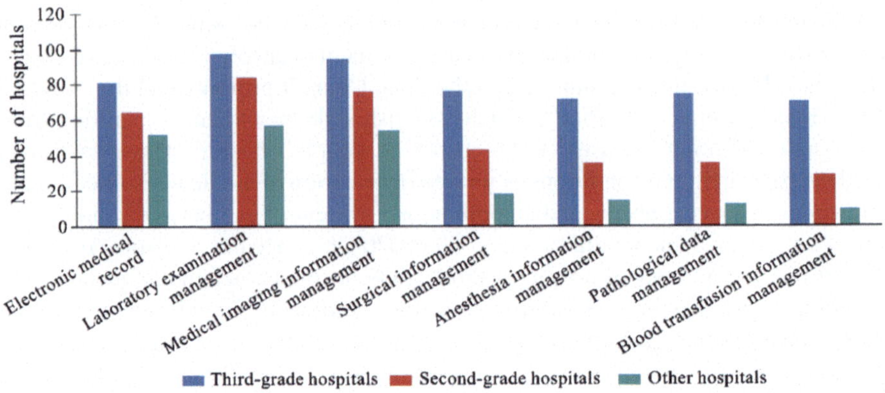

Fig. 6.18 Development of clinical medical applications in hospitals at all grade in China. *Data Source* Statistical Information Center of the National Health Commission, *Universal Health Informatization Survey Report—Regional Health Informatization and Hospital Informatization (2021)*

medical and health institutions such as hospitals, community health service centers (stations), and township-level health centers (village health clinics). The NHC regarded the grading evaluation of the application level of electronic medical record systems as one of the iconic indicators of the informatization of medical and health institutions. As of July 2021, a total of 118 hospitals in China had participated in the electronic medical record evaluation and were rated as the fifth-level and above medical and health institutions; 153 regional competent units and 503 s-grade and above hospitals had passed the interconnection level evaluation.

At present, China has been actively promoting regional medical and health informatization, using tiered diagnosis and treatment and data sharing to provide data sharing services to institutions and individuals such as medical service providers, health management institutions, patients, medical payers, and pharmaceutical product suppliers within a certain geographical scope. For example, Jiangsu Province has built a universal health information platform covering the entire province. As of May 2022, this platform had connected to 13 municipal and 96 county-level (or district-level) health information platforms in Jiangsu Province, connecting more than 20,000 medical and health institutions, and formulated and issued nearly 40 local health information standards, achieving standardized data collection and gathering 41.8 billion pieces of data. Based on the universal health information platform, Jiangsu Province has built a one-stop people-friendly service system called "Jiangsu Health Pass", which has provided 10 million health records for residents across the province; more than 10 business applications have been carried out, including comprehensive supervision of medical services, primary health supervision, maternal and child health management, and infectious disease monitoring and early warning.

At the same time, China has also been promoting the informatization of medical insurance payment services. In 2022, China built a unified national medical insurance information platform, connecting to about 400,000 designated medical institutions and 400,000 designated retail pharmacies, effectively covering all insured persons. The national medical insurance service platform has 280 million real-name users, covering more than 100 service functions. New applications of digital technology in auxiliary diagnosis, rehabilitation, distribution and transportation, medical robots and other aspects have been rapidly popularizing, and new forms such as Internet live interactive family parenting, online infant and toddler raising courses, and parent classes have been constantly emerging.

6.4.2 Internet Medical Application

Internet medical applications have been widely deployed, with offline and online medical care showing trend toward collaborative development. As of October 2020, there were more than 900 Internet hospitals in China, the telemedicine collaboration network had covered all prefecture-level cities, and more than 5500 hospitals at or above the second grade were able to provide online services. By the end of 2021, 64.6% of public hospitals at or above the second grade were able to provide telemedicine, and 54.5% had opened appointment diagnosis and treatment services. The service capacity of the health information platform of traditional Chinese medicine clinics continues to improve, and a total of 16,200 traditional Chinese medicine clinics have been connected to the platform. By the end of December 2020, the platform had received a total of 13 million patients, filled out nearly 1 million electronic medical records of traditional Chinese medicine, prescribed more than 870,000 prescriptions for treatment based on syndrome differentiation, and queried the knowledge base more than 1.03 million times. New health service models such as "Internet Plus" epidemic prevention science popularization, online consultation, remote consultation, and drug distribution have rapidly gained popularity.

During the COVID-19 pandemic in February 2020, more than 42,000 medical staff from all over the country rushed to Hubei Province to provide help. At the same time, more doctors took advantage of information technology to open up a "second battlefield" online, breaking through the bottleneck of medical service supply and effectively supporting the main battlefield in Hubei Province. For example, the medical team from Guangdong Province that provided pairing assistance[7] to Jingzhou City in Hubei Province collaborated with Jingzhou City to establish an Internet hospital. More than 1300 doctors from 15 medical institutions in Guangdong Province volunteered to participate,and over 100,000 visits were made in 18 days, providing sustained medical support for Jingzhou City. "Internet + Medical Health" has broken

[7] During the COVID-19 pandemic in 2020, China implemented a "pairing assistance" mechanism, where 16 provinces were designated by the State Council's joint prevention and control system to provide one-on-one support to prefecture-level cities in Hubei (excluding Wuhan), alleviating pressure on local medical resources.

the time and space limitations of the distribution of medical resources, and played an active role in the fight against the prevention and control of the COVID-19 pandemic by innovating service models, improving service efficiency, and reducing service costs.

5G smart emergency treatment has been deployed to improve the treatment efficiency. In March 2020, Hainan Province started the construction of 5G smart emergency platforms in Qionghai People's Hospital and Boao Central Hospital. From the first moment the emergency personnel receive the patient, the patient's basic vital signs data, condition assessment images, and emergency condition records are transmitted to the command center and emergency station in real time through the 5G network. Hospital experts can view patient testing and monitoring information and videos from the emergency ambulance, and provide remote guidance to the medical staff on the emergency ambulance through the remote consultation system and formulate rescue plans in advance (see Fig. 6.19). In 2020, Hainan Province started the construction of the "Capacity Improvement Project for Primary Medical and Health Institutions Based on the 5G-assisted IoTs". By April 2021, the project had realized the 5G smart upgrade of seven third-grade and class A hospitals, 18 county-level hospitals in cities and counties, 340 township-level medical institutions and 2693 village health clinics in the province.

5G remote surgery has begun operation. The 5G technology, characterized by high bandwidth and low latency, shortens the network latency to 10 ms, making remote surgery possible (see Fig. 6.20). In January 2019, the world's first remote animal surgery based on 5G network was successfully implemented in Fuzhou City, Fujian Province. The doctor and the "patient" (piglet) were 50 km apart. Through the 5G network environment built by China Unicom in Fujian Province, the doctor remotely controlled the surgical robot and successfully removed the piglet's liver lobule. In March 2019, China's first 5G-based remote human surgery—Parkinson disease "brain pacemaker" implantation surgery was completed in Beijing. This surgery was carried out through the 5G network, spanning nearly 3000 km, and successfully realized the Parkinson disease "brain pacemaker" implantation surgery between the First Medical Center of the Chinese People's Liberation Army General

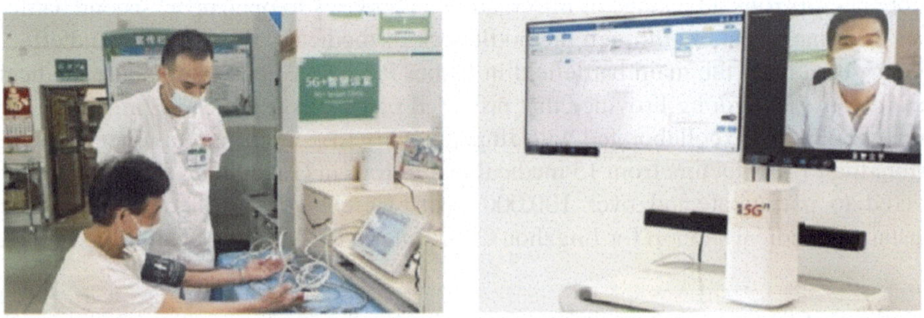

Fig. 6.19 5G remote emergency: Hainan Province's 5G smart emergency platform. *Image Source* People's Daily

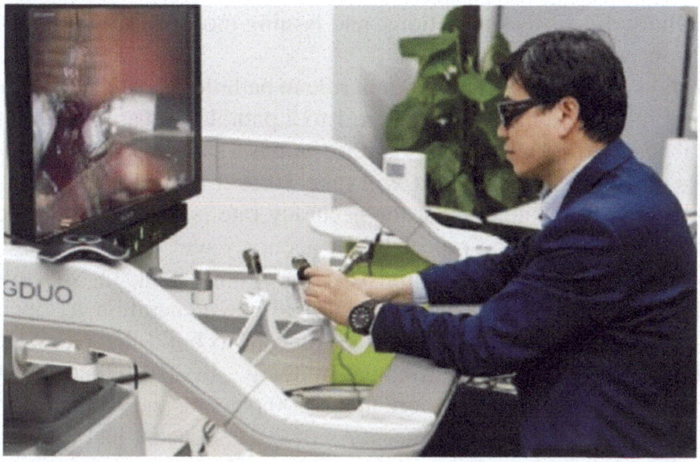

Fig. 6.20 5G remote surgery. *Image Source* People's Daily

Hospital in Beijing and one hospital in Hainan Province, realizing 5G remote surgery control.

6.4.3 Smart Medical Application

AI technology has been introduced in the assistance of diagnosis. For a long time, China's rural primary medical resources had been in a relatively disadvantaged position. The use of AI technology can provide auxiliary diagnosis suggestions for primary doctors and improve their diagnostic and treatment capabilities. "Smart Medical Assistant" is an artificial intelligence application for medical and health institutions promoted by the Health Commission of Anhui Province. Based on the medical knowledge graph, the application extracts and analyzes the electronic medical record information of the patient to provide doctors with a list of suspected diseases. "Smart Medical Assistant" has the monitoring and early warning capabilities for all 40 statutory communicable diseases, 15 key communicable diseases, and 6 syndromes. In 2017, this application passed the written examination of the national Medical Comprehensive Written Examination for Physician Qualification. "Smart Medical Assistant" also supports batch voice outbound calling, which can help primary doctors complete chronic disease follow-up, maternal and child health follow-up, notification, education, physical examination appointments and other tasks for contracted residents. As of October 2022, the "Smart Medical Assistant" had achieved full coverage and regular application in 104 districts and counties in Anhui Province, serving more than 30,000 primary doctors, assisting doctors in completion of more than 160 million electronic medical records, providing more than 430 million auxiliary diagnosis suggestions, and issuing more than 130 million rational drug quality inspection prescriptions.

AI also plays an increasingly important role in pathological diagnosis. According to the current medical standards in China, 1 to 2 pathologists are required for every 100 medical beds. Based on this calculation, there is a shortage of 80,000 to 100,000 pathologists in China. In general, it will take 30 to 40 min for a human pathologist to read each film manually, but the accuracy rate is only 60%. AI technology can help doctors improve the repeatability, accuracy and efficiency of diagnosis, and alleviate the shortage of pathologists to a certain extent. In November 2017, the MOST announced the first batch of national new-generation AI open innovation platforms, and specifically designated Tencent to build the "New-Generation AI Open Innovation Platform for Medical Imaging". In 2018, Tencent established an AI medical center, applied for more than 1500 patents, and published more than 300 papers at top academic conferences. As of 2019, Tencent had cooperated with more than 100 third-grade and class A hospitals in China, processed more than 270 million images in medical imaging, and developed AI imaging products such as early esophageal cancer screening, early lung cancer screening, diabetic retinopathy detection, colorectal cancer screening, early breast cancer screening and early cervical cancer screening; the AI auxiliary/guidance system had been introduced into 300 hospitals and 4.7 million accurate guidance had been carried out.

6.5 Summary

The Chinese government attaches great importance to improving the living standards and quality of the people. In recent years, China has given full play to its advantages in informatization and established a comprehensive and multi-level digital social service system, which has greatly expanded the coverage of public services and significantly enhanced the level of equalization of basic public services. Through the construction of smart city and digital village, urban and rural public services have achieved coordinated optimization, making people's lives more convenient and better. The in-depth advance-ment of smart education and smart healthcare has effectively promoted educational equity and the rational distribution of medical resources.

China regards smart city as an advanced stage of urban informatization construction and actively uses informatization means to address development challenges in densely populated cities. In terms of urban traffic control, many cities, such as Shenzhen, have built intelligent traffic command and control centers, greatly improving traffic efficiency. In urban comprehensive governance, China pioneered the concept of the "City Brain". Practices in places like Hangzhou have achieved fruitful results and effectively solved urban operation problems. In terms of urban digital governance, cities such as Shanghai have explored the construction of digital twin city systems, making urban management more efficient and intelligent.

China attaches great importance to coordinated urban–rural development and regards the construction of digital villages as a key path to rural revitalization. In the construction of rural network infrastructure, China has maintained continuous investment over the years. Through projects such as the "Village-to-Village" project, the Internet-based Poverty Alleviation Action Plan, China has overcome numerous difficulties and achieved full coverage of rural network infrastructure, greatly narrowing the "digital divide" between urban and rural areas. As to the rural digital economy, the government actively promotes the development of rural e-commerce. Innovative models such as Taobao Villages have driven the sales of agricultural products and the employment of farmers, providing new ideas for global rural economic development and poverty reduction. In addition, China has also used digital methods to improve the level of rural public services, enhance the level of rural grass-roots comprehensive governance, and enrich the cultural lives of farmers.

China regards education informatization as an important way to address the imbalance of educational resources and actively promotes the digital transformation of education. In the construction of digital campus facilities, the network conditions of schools have been significantly improved, high-quality educational resources are being effectively shared, and a complete digital campus standard system has been established. Digital course platforms are constantly being improved, covering primary and secondary education as well as higher education. They played an important role during the pandemic, enabling continuous learning despite school closures. Internet-based education applications represented by the "Three Classrooms" have promoted educational equity in basic education and effectively addressed the shortage of educational resources in remote rural areas. AI general education, represented by IT education and robot innovation, is being gradually promoted in primary and secondary schools, providing support for cultivating a scientific research talent team for China in the future.

China takes medical informatization as an important means to address the uneven distribution of medical resources. In terms of digital medical infrastructure, China has established a basic medical security system covering the entire population, formulated standards for hospital informatization, and significantly improved the modernization level of medical institutions at all levels and of various types. In the application of Internet-based medical services, the ICT applications such as 5G have achieved breakthroughs in remote first aid and remote surgery, making up for the development gap in medical levels between regions. In the application of new technologies, the AI applications in auxiliary diagnosis and pathological diagnosis are becoming more and more mature. These measures improved the medical level at the grass-roots level, alleviated the healthcare workforce shortage of medical resources, and contributed Chinese wisdom to global medical development.

It is worth noting that during the construction of a "Smart Society", the advantages of socialism with Chinese characteristics have played a crucial role in the face of many difficulties and challenges. Take rural network construction as an example, although facing problems such as high construction investment, high maintenance costs, and low investment returns, Chinese state-owned enterprises, proceeding from national strategies and the interests of the people, have bravely taken on heavy responsibilities and made continuous investments. They have successfully extended information networks to every remote village, building the world's most extensive rural network infrastructure. This has provided solid support for China to win the battle against absolute poverty, demonstrating the strong advantage of the socialism with Chinese characteristics in pooling resources to accomplish major tasks and offering a successful example for the construction of a global digital society.

Chapter 7
Cyberspace Governance

7.1 Demands and Challenges

7.1.1 Overview of Cyberspace

The concept of cyberspace was initially introduced by Western scholars and has gradually gained global recognition. In 1984, American science fiction writer William Gibson first proposed the concept of cyberspace in his novel Neuromancer, referring to the space formed by the global computer network. Major countries in the world have different definitions of cyberspace. For example, the United States (US) describes cyberspace as "an interdependent network of mutual trust composed of information technology infrastructure, including the Internet, telecommunications networks, computer systems, etc., as well as a virtual environment for information and human interaction"; Germany defines cyberspace as "including all information infrastructure that can be accessed through the Internet across territorial boundaries"; the United Kingdom (UK) defines cyberspace as "a human–computer interaction field composed of digital networks and used to store, modify, and transmit information, including the Internet and other information systems used to support business, infrastructure and services"; China's "National Cyberspace Security Strategy" points out that cyberspace is "composed of the Internet, communications networks, computer systems, automatic control systems, digital devices and the applications, services and data they carry".

Cyberspace has become the overarching term for the digital society constructed by information and network technologies and products. The omnipresence of the Internet makes a secure, stable, and prosperous cyberspace increasingly significant for national and global peace and development. Cyberspace governance encompasses various aspects, including cyberspace information security, legal construction in cyberspace, and regulation of the digital economy. In comparison to governance in traditional human societies, cyberspace governance exhibits four distinctive characteristics.

W. Liu et al. (eds.), *Digital China*, Introduction to China's S&T Innovation,
https://doi.org/10.1007/978-981-95-6240-4_7

Firstly, cyberspace governance is fundamental. With the rapid development of global informatization, network information systems have become key platforms and nerve center for national politics, economy, culture, and society. Nearly all the activities in social production and life are increasingly inseparable from network information systems, including state organs and various enterprises and institutions fulfilling their social management responsibilities, providing social services, and maintaining their normal operations.

Secondly, cyberspace governance is open. The Internet is inherently open, and information exchange in cyberspace transcends geographical boundaries and is not constrained by space. This characteristic also means that cyberspace security has no obvious boundaries. In this sense, cyberspace security is not only the responsibility of a country, but also the common responsibility of the international community.

Thirdly, cyberspace governance is systematic. Cyberspace governance encompasses everything from the hardware of network construction and the software managing and controlling networks to network services and applications. The higher the degree of informatization, the greater the number of distributed nodes. Any node in the network can impact cyberspace security, akin to the weakest-link principle. Complex network systems require a focus on the security levels of the most vulnerable elements.

Finally, cyberspace governance is strategic. As cybersecurity and informatization continue to play an increasingly crucial role in national security and development, they have risen to the level of major national strategies. China needs to strengthen top-level design and strategic coordination. Cyberspace security represents a strategic mindset and deployment in the era of the Internet.

7.1.2 Cyberspace Security and Governance

Cyberspace security represents a novel form of societal security, encompassing the security of governments, social organizations, and individuals on a global scale. It primarily involves three aspects: the security of computer systems, the security of programs and data within the systems, and the security of network connections. The security of computer systems, forming the physical foundation of cyberspace security, not only supports the normal functioning of the network, but also accepts services, serving as a critical guarantee for the network's normal operation. The security of programs and data within the systems is essential for ensuring the stable and reliable operation of applications and protecting software data from damage and theft. The security of network connections refers to the protection of information during the process of network connection, such as protecting the transmission path and specific content from malicious alteration, defending against the intrusion of Trojans and viruses, and more.

Cyberspace has gradually become a crucial domain for social life, economic interactions, political maneuvering, and cultural integration. However, it has not fundamentally altered the basic legal order of human society. The actions of individuals and groups in both physical space and virtual cyberspace constitute their complete behavior in human society, requiring protection and supervision under the rule of law. The continuous interaction and collaboration of humans in the virtual cyberspace have promoted the formation of the cyber society. The concept and scope of cyber society governance are continuously evolving. Representative issues include the protection of user digital rights, governance of online behavior, technical algorithm governance, digital economic regulation, and more. Specifically:

First, the connotation of user digital rights is relatively broad, usually including the right of every citizen to access the Internet normally, the right to conduct relevant activities on the Internet, and the right to protect the privacy of personal digital information. Internet access services have become public services like ordinary postal services. Civil activities carried out by individuals on the Internet are also legally binding, such as payment and signature, which require a sound personal identity authentication system to ensure. Individuals participation in online activities, such as online shopping and online medical consultation, also involves personal privacy and other sensitive information, which requires a sound personal information protection mechanism.

Second, there are also many illicit activities and illegal behaviors in cyberspace, which need to be governed in line with the existing laws and regulations. Typical illegal online behaviors involve the dissemination of illegal content, including editing and disseminating infringing digital content, disseminating violent and reactionary content, and disseminating false information. In addition, using the Internet to engage in illegal activities will also damage the rights and interests of the general public. For example, illegal activities such as online fraud, online gambling, online pyramid schemes, and online illegal fundraising will cause significant losses to the victim's property and affect the normal order of life.

Third, the technology and algorithm governance of cyberspace faces multi-dimensional challenges. Artifical intelligence (AI) technology, dominated by machine learning algorithms, has become the core driving force of major online platforms. However, its structural flaws in terms of fairness, transparency, and security have led to frequent occurrences of irregular algorithm applications. These technological risks have gradually evolved into complex challenges for cyber society governance, affecting many aspects such as social fairness and justice, the legitimate rights and interests of Internet users, and the security of virtual property.

Fourth, competition in the online market during the digital economy era has brought many new problems. For example, super platforms have become digital economic infrastructure formed in a highly digital market environment, and the operation and supervision of platforms are difficult to manage through traditional market supervision methods. For another example, online commodity transactions involve not only online links such as buying and selling, and after-sales service, but also offline links such as express logistics, which are difficult to track and supervise through traditional channels.

Cyberspace is not beyond the reach of the law. Safeguarding the lawful rights of individuals and groups in cyberspace and maintaining the normal operation of the cyberspace are new challenges for every government. Countries worldwide are taking measures, including legislative actions, to govern behaviors in the cyber society. Taking 2020 for instance, in January, the US released the "Guidance for Regulation of AI Applications", emphasizing flexible regulatory approaches to avoid hindering innovation; in February, the European Commission released the "White Paper on Artificial Intelligence: a European approach to excellence and trust" and proposed to build a "regulatory framework for AI"; in June, the European Commission assessed the "General Data Protection Regulation", and established a personal data usage model based on principle that "processing personal data is generally prohibited, unless it is expressly allowed by law"; in December, the European Commission unveiled drafts of the "Digital Services Act" and the "Digital Markets Act", requiring that technology companies must not utilize competitors data to compete against them and should ensure fair product displays on their platforms.

7.2 Cyberspace Legal System

With the development of the Internet, China's cyber legislation has gone through a development process from nothing to something, from few to many, from points to lines, from lines to planes, and from planes to volumes.

The first stage ran from 1994 to 1999, which represented the access popularization of the Internet in China. The number of Internet users and devices increased steadily. The cyber legislation at this stage mainly focused on network infrastructure security, namely computer system security and cybersecurity.

The second stage ran from 2000 to 2011, marking the extensive application of the Internet, a period when personal computers served as the primary terminal devices for Internet access. As the number of computers gradually increased and Internet connection services became more affordable, Internet access became increasingly common, and network information services developed rapidly. During this stage, cyber legislation shifted its focus to network information services and content management.

The third stage began from 2012, which corresponded to the widespread adoption of mobile Internet, with smartphones becoming the predominant terminal devices for accessing the Internet. Cyber legislation at this stage gradually tends to comprehensively cover the network information services, information development, cybersecurity, and other aspects of comprehensive Internet governance.

As of 2023, China had established over 140 cyberspace-related laws, forming a comprehensive legislative framework for digital governance. Anchored in the Constitution and strengthened by multi-tiered legal instruments—including laws, administrative regulations, departmental rules and local regulations, and local administrative rules—this framework extends from the traditional legislation, addresses the special domains of cyberspace, such as online content construction and governance, cybersecurity and informatization development, and provides institutional bedrock for advancing China's strategic cyberpower objectives. The overview of China's cyber legislation is shown in Table 7.1.

7.2.1 Protection of Personal Rights

Safeguarding citizens freedom and confidentiality of communications. The protection of freedom and confidentiality of communications is a prerequisite for ensuring that citizens can express their demands and ideas independently in cyberspace. As early as 1997, the State Council formulated the "Administrative Measures for the Security Protection of International Internet Connection of Computer Information Networks" to implement the protection of the basic rights of freedom and confidentiality of communications in the Constitution; the "Telecommunications Regulations of the People's Republic of China" promulgated in 2000 stipulates that the freedom of telecommunications users to use telecommunications in accordance with the law and the confidentiality of communications are protected by law; the "Radio Management Regulations of the People's Republic of China" revised in 2016 further strengthens the protection of confidentiality of communications in the field of radio, realizing the all-round protection of this basic right in cyberspace.

Protecting personal information rights and interests. Through civil law, criminal law, and special legislation, a legal barrier for the protection of the entire chain of personal information rights and interests has been established. The Amendment IX to the "Criminal Law of the People's Republic of China" adopted in 2015 makes more detailed provisions on the crime of illegally obtaining citizens personal information and strengthens the criminal law protection of personal information. In the special network legislation, the "Decision of the Standing Committee of the National People's Congress on Strengthening the Protection of Network Information" adopted in 2012 clearly protects electronic information that can identify citizens personal identities and involves citizens personal privacy; the "Cybersecurity Law of the People's Republic of China" promulgated in 2016 has further improved the rules for personal information protection; the "Personal Information Protection Law of the People's Republic of China" promulgated in 2021 refines and improves the principles of personal information protection and the rules for personal information processing, regulates the activities of state organs in handling personal information in accordance with the law, grants personal information subjects a number of rights, and strengthens the obligations of personal information processors.

Preserving the property safety of citizens. China continues to increase legislative protection efforts to curb the use of the Internet to infringe on property rights and interests. The "Electronic Commerce Law of the People's Republic of China" promulgated in 2018 stipulates that "the goods sold or services provided by e-commerce operators shall comply with the requirements for protecting personal and property safety and environmental protection requirements, and shall not sell or provide goods

Table 7.1 Overview of China's cyber legislation

Types	Examples
Laws	Electronic Commerce Law of the People's Republic of China
	Electronic Signature Law of the People's Republic of China
	Cybersecurity Law of the People's Republic of China
	Data Security Law of the People's Republic of China
	Personal Information Protection Law of the People's Republic of China
	Law on Combating Telecom and Online Fraud of the People's Republic of China
Administrative regulations	Regulations on the Protection of Computer Information Systems of the People's Republic of China
	Regulations on Computer Software Protection
	Administrative Measures on Internet Information Services
	Telecommunications Regulations of the People's Republic of China
	Regulations on the Administration of Foreign Investment in Chinese Telecommunications Businesses
	Regulations on the Protection of the Right of Communication Through Information Networks
	Regulations on the Security and Protection of Critical Information Infrastructure
Departmental rules	Regulations on the Protection of Children's Online Personal Information
	China Internet Domain Name Regulations
	Measures on the Supervision and Administration of Online Transactions
	Provisions on the Administration of Internet News and Information Services
	Regulations on the Governance of Internet Information Content
	Regulations on the Management of Algorithmic Recommendations for Internet Information Services
Local regulations	Guangdong Provincial Digital Economy Promotion Act
	Zhejiang Provincial Digital Economy Promotion Act
	Hebei Provincial Regulations on Information Technology Development
	Guizhou Provincial Regulations on Government Data Sharing
	Shanghai Municipal Data Regulations
Local administration rules	Measures of Guangdong Province on Public Data Management
	Measures of Anhui Province on the Management of Data and Resources for Government Affairs

(continued)

Table 7.1 (continued)

Types	Examples
	Measures of Jiangxi Province on the Protection of Computer Information System Security
	Interim Measures of Hangzhou City on the Administration of Online Transactions

or services prohibited by laws and administrative regulations". The "Civil Code of the People's Republic of China" clearly states that the use of the Internet to infringe on the property rights and interests of others shall bear corresponding legal responsibilities. The "Law on Combating Telecom and Online Fraud of the People's Republic of China" promulgated in 2022 provides strong legal support for combating telecom and online fraud activities and effectively safeguards the property rights and interests of the people.

Ensuring the digital rights for special groups. Through multi-level and multi-dimensional legislation, the digital divide of special groups such as minors, the elderly, and the disabled can be bridged, so that they can integrate into the digital society more equally and widely, and enjoy the dividends of the digital age. The "Cybersecurity Law of the People's Republic of China" stipulates that "the state supports the research and development of network products and services that are conducive to the healthy growth of minors, punishes the use of the Internet to engage in activities that endanger the physical and mental health of minors in accordance with the law, and provides a safe and healthy network environment for minors." The "Regulations on the Protection of Children's Online Personal Information" issued in 2019 gives priority to the protection of children's personal information rights and interests. The "Minors Protection Law of the People's Republic of China" revised in 2020 makes special provisions for strengthening the Internet literacy education of minors, strengthening the supervision of Internet content for minors, strengthening the protection of minors personal information and the prevention and treatment of Internet addiction, etc., to protect the legitimate rights and interests of minors on the Internet. The "Data Security Law of the People's Republic of China" issued in 2021 requires that intelligent public services should fully consider the needs of the elderly and the disabled, and avoid obstacles to the daily lives of them.

7.2.2 Digital Economy Regulation

Continuously improving data infrastructure system. China maintains digital market order, and regulates new digital economic formats and models, providing a good institutional foundation for the healthy development of the digital economy and helping the economy shift from high-speed growth to high-quality development. China pays attention to the role of data as a basic resource and an innovation engine. The "Data Security Law of the People's Republic of China" makes relevant provisions for implementing the big data strategy, supporting data-related technology research and development and commercial innovation, promoting the construction of data-related standard systems, and cultivating data trading markets, so as to improve the level of data development and utilization and promote the development of the digital economy with data as the key element.

Clarifying operational systems for the digital market. China adheres to the legal and standardized development of the digital market, resolutely opposes monopoly and unfair competition, improves digital rules, and effectively maintains a fair and competitive market environment. The "Electronic Commerce Law of the People's Republic of China" comprehensively regulates e-commerce business operations, clarifies the responsibilities of e-commerce platform operators and operators within the platform, and requires e-commerce operators with market dominance not to abuse their market dominance to exclude or restrict competition, and maintain a fair market competition order. The "Law of the People's Republic of China on the Protection of Consumer Rights and Interests" revised in 2013 established systems such as "seven-day no-reason return" for online shopping, and strengthened the main responsibility of online operators for consumer rights protection. The "Anti-Unfair Competition Law of the People's Republic of China" revised in 2017 added a special article on the Internet, prohibiting the use of technical means for unfair competition. The "Measures on the Supervision and Administration of Online Transactions" formulated in 2021 refines the relevant provisions of the "Electronic Commerce Law of the People's Republic of China" and further improves the online transaction supervision system.

Regulating new business formats and models in the digital economy. New business formats and models of digital economy in China are emerging rapidly, bringing tremendous impetus and potential to economic and social development while also posing new challenges to social governance and industrial development. The "Civil Code of the People's Republic of China" improves the rules for the conclusion and performance of electronic contracts, brings data and network virtual property into the scope of legal protection, and promotes the development of digital economy. The "Interim Measures for the Administration of Online Taxi Booking Services", the "Regulations on the Management of Algorithmic Recommendations for Internet Information Services", the "Provisions on the Administration of Blockchain Information Services", the "Interim Measures for the Administration of Business Activities of Online Lending Information Intermediaries", and the "Interim Provisions on the Administration of Online Tourism Services" regulate new technologies and formats of business such as online taxis, algorithms, blockchain, Internet finance, and online tourism, and enrich the legal basis for the governance of various fields of "Internet Plus".

7.2.3 Network Information Security

Establishing cybersecurity rules. In 1994, China issued the "Regulations on the Protection of Computer Information Systems of the People's Republic of China", which established the protection system and security supervision system of the computer information system security; in 2000, the "Decision of the Standing Committee of the National People's Congress on Ensuring Internet Security" was passed, which divided Internet security into Internet operation security and Internet information security, and established a cybersecurity responsibility system framework of civil liability, administrative liability, and criminal liability; on November 7, 2016, the "Cybersecurity Law of the People's Republic of China" was passed, which is China's first basic law to comprehensively regulate issues related to cybersecurity management. The "Cybersecurity Law of the People's Republic of China" clarifies the systems for ensuring network operation security, network products and services security, network data security, and network information security. Laws and regulations such as the "Measures on Cybersecurity Review" and the "Regulations on the Management of Security Loopholes of Network Products" further refine the relevant systems of the "Cybersecurity Law of the People's Republic of China". In May 2019, the "Information Security Technology—Basic Requirements for Cybersecurity Level Protection" was officially released, making the official entry of China's cybersecurity level protection work into the 2.0 era.

Ensuring security of critical information infrastructure. Critical information infrastructure is the nerve center of economic and social operations, and is of great significance for maintaining national cyber sovereignty and national security, ensuring the healthy development of the economy and society, and safeguarding public interests and the legitimate rights and interests of citizens. The "Regulations on the Security and Protection of Critical Information Infrastructure" passed in 2021 has improved the critical information infrastructure identification mechanism, and made specific provisions for operators of critical information infrastructure to implement cybersecurity responsibilities, establish and improve cybersecurity protection systems, set up special security management institutions, conduct security monitoring and risk assessments, and standardize network products and services procurement activities, providing a legal basis for accelerating the improvement of critical information infrastructure security protection capabilities.

Constructing a legal system for data security management. Based on the actual data security work and focusing on prominent issues in the field of data security, China has strengthened data security protection through legislation to enhance the country's data security protection capabilities. The "Data Security Law of the People's Republic of China" was officially issued in June 2021, achieving a "breakthrough" from scratch in the data governance system and becoming a basic law in the field of China's data security. The "Data Security Law of the People's Republic of China" clearly stipulates the establishment and improvement of systems such as data categorization and classification protection, risk monitoring and early warning, emergency response, and data security review; it also contains measures to facilitate data security and development and provisions for the security and openness of government data.

7.2.4 Network Ecosystem Norms

The cyberspace serves as a collective spiritual haven for hundreds of millions of people. A cyberspace characterized by a clear and pure atmosphere and a thriving ecosystem, embodies the aspirations of the populace for an ideal online home. Out of a strong sense of responsibility toward society and the people, China has introduced laws and regulations for cyberspace governance, focusing primarily on regulating online information content and persistently purifying the online environment.

Standardizing the order of online information dissemination. Facing with the global problem of online information governance, China has promulgated laws and regulations such as the "Civil Code of the People's Republic of China", the "Cybersecurity Law of the People's Republic of China", and the "Administrative Measures on Internet Information Services" to clarify the norms for the dissemination of online information content and the responsibilities of relevant entities, providing a legal basis for the governance of illegal acts that endanger national security, damage public interests, and infringe upon the legitimate rights and interests of others.

Building legal instruments to combat terrorism in cyberspace. China stands firm to against the threat of cyberterrorism in accordance with the law. The "Criminal Law of the People's Republic of China", the "Criminal Procedure Law of the People's Republic of China", the "Anti-Money Laundering Law of the People's Republic of China" and other laws have made provisions for the criminal liability of terrorist crimes, the litigation procedures for punishing terrorist crimes, and the monitoring of terrorist funds. The "Anti-Terrorism Law of the People's Republic of China", passed in 2015, has made special provisions on the objects, measures and mechanisms of cyberterrorism.

7.3 Cybersecurity Protection

7.3.1 Cybersecurity Technology Standards

China has strengthened research on new technologies, developed and deployed a series of security products and system platforms, and overcome a number of major technical challenges in the field of information security. China has independently developed the ZUC[1] stream cipher algorithm, which has encryption strength comparable to that of international mainstream cipher algorithms. Significant progress has been made in the joint efforts to tackle key problems in the field of secure and reliable basic software and hardware. Self-developed security operating systems have progressed from "basic availability" to "availability". Domestic databases are widely used in state organs and important fields. Domestic computers based on domestic

[1] ZUC is named after Zu Chongzhi, a renowned Chinese mathematician and astronomer of the 5th century AD.

basic software and hardware have been tried out in some party and government agencies and key fields. The industrialization of security protection systems for industrial control systems have made positive progress. In August 2010, the TePA (Tri-element Peer Authentication) proposed by China was officially approved by the International Organization for Standardization (ISO), becoming the first international standard approved in the field of basic common technologies for cybersecurity.

China actively promotes cybersecurity standardization and builds a sound national cybersecurity standard system. In 2020, the National Technical Committee on Information Security Standardization promoted the release of 53 national cybersecurity standards and more than 10 international standards led and co-developed by China. In 2021, China made new breakthroughs in the international standardization of commercial cryptography. Its independently developed SM9 identification cipher algorithm was officially released as an ISO/IEC (International Organization for Standardization/International Electrotechnical Commission) international standard, and the SM4 block cipher algorithm entered the ISO/IEC official release stage. As of June 2021, China had issued a total of 323 national cybersecurity standards, which play a fundamental, normative and leading role in supporting national cybersecurity work and safeguarding the vital interests of the majority of Internet users.

7.3.2 Cybersecurity Protection Forces

China has established a number of professional cybersecurity institutions, which have effectively improved China's cybersecurity monitoring and emergency response capabilities. In 1996, the National Computer Virus Emergency Response Center was established to mobilize domestic anti-virus forces to quickly discover and deal with computer virus epidemics and network attacks, providing important technical support for the country to formulate computer virus prevention and control policies, regulations and standards. In 1998, the China Information Technology Security Evaluation Center was established to carry out security vulnerability analysis and information notification of information technology products and systems. In 2001, the National Computer Network Emergency Response Technical Team/Coordination Center of China (CNCERT/CC) was established to carry out prevention, discovery, early warning, and coordination and disposal of Internet cybersecurity incidents. In 2004, the National Cybersecurity and Information Security Notification Center was established as a national cybersecurity and information security notification outlet to report and notify information on China's network information security. In 2005, the National Information Technology Security Research Center was established to mainly conduct the security analysis of and research on information technology products and systems and provide the information security assurance of national basic information networks and important information systems.

At present, China has developed the technical capabilities for information content supervision, such as identifying harmful information on the Internet and early detection of emergencies, the technical capabilities for overall confrontation in cyberspace, such as emergency response and coordinated handling of cybersecurity incidents and in-depth protection of information systems, the technical capabilities for information resource protection, such as centralized network access for government departments and network identity authentication management, and the technical capabilities for security support in new technologies and new applications such as cloud computing and the Internet of Things (IoT), forming a relatively sound cybersecurity protection capability system.

In 2020, the CNCERT/CC coordinated and handled about 103,000 cybersecurity incidents of various types, a decrease of 4.2% compared with the previous year. From 2016 to 2020, the number of hosts infected with computer malicious programs in China continued to decline and remained at a low infection level, with an average annual decreasing rate of 25.1%. Through the China Cybersecurity Threat Governance Alliance, 10 domestic browser manufacturers has cooperated in a collaborative defense pilot to provide prompts and interception when users visit phishing websites. The number of prompts and interceptions reached 390 million times in 2020. The CNCERT/CC continues to carry out governance work on network resources used for distributed denial of service (DDoS) attacks. Consequently, such exploitable attack resources within China are continuously declining, and DDoS attacks have been curbed at the source. According to statistics, the annual decreasing rate of DDoS attacks in China in 2020 was 16.16%, the total attack traffic decreased by 19.67% year-on-year, and the global share of botnet control terminals in China steadily decreased to 2.05%.

7.3.3 Development of Cybersecurity Industry

China's cybersecurity industry has evolved from non-existence to establishment, and expanded from small-scale to large-scale. The industrial structure has gradually become complete, and the industrial agglomeration effect has begun to emerge, playing an important role in promoting the healthy development of the Internet. Projects such as the "Information Security Special Project" of the National Development and Reform Commission (NDRC), the "National Science and Technology Support Program" of the Ministry of Science and Technology (MOST), and the "Electronic Information Industry Development Fund Project" of the Ministry of Industry and Information Technology (MIIT) effectively promote the industrialization of cybersecurity. In September 2019, the MIIT publicly solicited opinions on the "Guidance on Promoting the Development of the Cybersecurity Industry (Draft for Comments)", which is committed to strengthening the emphasis on cybersecurity, increasing the application scale of cybersecurity products, and actively innovating cybersecurity services models.

China has developed cybersecurity products covering various fields, including physical security, communications security, data security, application security, security management platform, and new application security, forming a relatively complete information security industry chain from security chips, network and border security products, data security products, application security products to security services. A fully controllable system of independent research and development, production, upgrading and maintenance from basic software and hardware to application software has basically formed. In terms of basic hardware, domestic chip products such as processors, switching chips, and display chips have approached the level of mainstream foreign products, and have the conditions for systematic deployment. In terms of basic software, domestic operating systems and domestic database technologies have become mature and covered fields such as servers, desktops, mobile devices, and embedded systems. In terms of application software, China's independently controllable software for office automation, enterprise management, and industrial application system has been applied in communications, military, aviation, aerospace, and other fields with high-precision technologies and high real-time requirements.

In 2019, the scale of China's cybersecurity industry reached 156.36 billion yuan, with an annual growth rate of 17.1%, significantly higher than the international average growth rate of 9.11%, maintaining a healthy development trend. By the end of November 2019, there were 23 listed cybersecurity enterprises in China, more than 100 venture capital institutions invested in the field of cybersecurity, and more than 150 innovative and entrepreneurial enterprises were gathered. According to the "2020 China Cybersecurity Industry Statistical Report" released by the Cyber Security Association of China in June 2020, the total revenue of China's cybersecurity technology, products and services in 2019 was about 52.31 billion yuan, with an annual growth rate of 25.37%; the number of employees in cybersecurity enterprises was about 100,000.

7.4 Cyberspace Governance

7.4.1 Public Environmental Governance

China has established and improved the basic legal environment for the Internet, and has successively issued a series of laws and regulations related to the Internet, covering the management of Internet infrastructure resources, information dissemination norms, market order norms, information security guarantee, and other aspects, and has made provisions for the responsibilities and obligations of Internet access service providers, Internet information service providers, government management departments, and Internet users. According to their duties, relevant departments have issued a series of departmental regulations and normative documents in the fields of Internet news and information services, news publishing, audio-visual programs, online games, online culture, online medicines, online copyrights, online transactions, electronic certification, network domain names, Internet access, and system security and confidentiality.

In terms of the management of Internet infrastructure resources, China has strengthened the management of domain names, IP addresses and website registration and filing, access services and others, in accordance with the law. On December 20, 2004, the "Regulations on the Administration of Internet Domain Names in China" was officially implemented, which ensured the safe and reliable operation of China's Internet domain name system and standardized the domain name related to registration and management. In February 2005, the Ministry of Information Industry launched the centralized filing of Internet sites nationwide, and gradually established three basic databases including Internet content provider (ICP) filing information, IP address usage information, and domain name information. As of 2015, the filing rate of Chinese websites had increased to 99.98%, and the accuracy rate of the filing subject information had reached 84.7%.

In terms of user account name management, China issued the "Regulations on the Administration of Business Premises of Internet Access Service" in September 2002 to strengthen the management of Internet access service business premises and regulate the business behavior of operators. In 2012, China began to authenticate the real identity information of users registering accounts for Internet information service; from September 1, 2013, new fixed-line phones and mobile phones (including wireless network cards) were required to register their real identity information. On January 22, 2021, the Cyberspace Administration of China (CAC) issued the revised "Regulations on the Management of Internet User Public Account Information Services", which made the management of public accounts more refined and precise.

7.4.2 *Personal Privacy Protection*

Data security is closely tied to personal safety. Once a large amount of personal privacy of citizens, such as personal identity, family, economic status, hobbies, and irreversible biometric information such as face, fingerprint, and DNA are maliciously leaked, the security risks cannot be underestimated. Based on this, China has successively formulated laws and regulations such as the "Cybersecurity Law of the People's Republic of China", the "Electronic Commerce Law of the People's Republic of China", and the "Civil Code of the People's Republic of China". The "Data Security Law of the People's Republic of China" implemented in September 2021 and the "Personal Information Protection Law of the People's Republic of China" implemented in November 2021 constitute the other two pillars of China's continuous efforts to strengthen personal information security and privacy protection.

Since 2020, China has widely promoted applications for epidemiological surveys and the protection of public health and safety. It is particularly important to protect the personal privacy data in these applications. On March 6, 2020, China issued a new version of the "Information Security Technology—Personal Information Security Specification", which clearly stated that "trajectory" belongs to "personal sensitive information", and the collection of such information must adhere to the "minimum necessary principle" and the "legitimacy principle", with authorization only waived in 11 exceptional situations "directly related to public safety, public health, and significant public interests".

7.4.3 Individual Rights Protection

Mobile Internet applications have become a focal point for the protection of personal information. In March 2021, the Cyberspace Administration of China and other departments jointly issued the "Regulations on the Scope of Necessary Personal Information for Common Types of Mobile Internet Applications", which clarified the basic functions of 39 basic types of applications and the specific types and usage requirements of personal information collected to ensure their normal operation. In April 2021, the MIIT, together with the Ministry of Public Security (MPS) and the State Administration for Market Regulation (SAMR), drafted and promulgated the "Interim Regulations on the Protection and Management of Personal Information of Mobile Internet Applications (Draft for Comments)", which detailed the identification standards of "informed consent" and "minimum necessary", and promoted the effective solution to the problem of collection and use of user's personal information in practice such as unclear rules, purpose, method, and scope.

In terms of the protection of minors online content, the revised "Law of the People's Republic of China on the Protection of Minors" officially came into effect on June 1, 2021, opening a new chapter in the protection of minors online content. Relevant departments have successively issued normative documents and deployed special actions related to the protection of minors online content. A series of special actions have been jointly carried out across the country to combat pornography and obscenity on the Internet and mobile media, rectify vulgarity on the Internet, rectify cyberbullying and violence, rectify unlicensed Internet cafes, and clean up and rectify online games. News websites and commercial websites have been organized to self-examine and self-correct and clean up all kinds of illegal and harmful information, and the online environment has been continuously purified.

In terms of supporting the elderly in using the Internet, in December 2020, the MIIT issued the "Special Action Plan for Elderly-Oriented and Accessibility Renovation of Internet Application" to deploy relevant products and services measures. In April 2021, the MIIT issued the "General Design Specifications for Elderly-Oriented Mobile Internet Applications" and the "General Design Specifications for Elderly-Oriented Internet Websites", which clearly prohibit the inclusion of advertising content and plug-ins in the elderly-oriented interface misleading buttons that induce downloads and payments. In June 2021, the MIIT officially issued three standards, namely "Technical Requirements of Mobile Terminal Suitability for Elderly Person", "Test Methods of Mobile Terminal Suitability for Elderly Person", and "Technical Requirements for Elderly-Oriented Consideration of Smart Television", focusing on solving the various difficulties encountered by elderly person when using smart terminal products.

In terms of protecting the rights and interests of workers in the digital economy, China pays attention to protecting the interests of the majority of workers, consumers and other groups, maintaining social fairness and justice, and ensuring the sustained, stable and healthy development of the digital economy. On July 26, 2021, the SAMR, the Cyberspace Administration of China (CAC) and five other departments jointly issued the "Guidelines on Implementing Responsibilities of Online Food Delivery Platforms and Effectively Safeguarding the Rights of Delivery Workers" in the form of a normative document, which puts forward comprehensive and specific requirements for protecting the legitimate rights and interests of delivery drivers, and requires that the legitimate rights and interests of delivery drivers be effectively protected from 10 aspects, including labor income, assessment system, dispatch mechanism, and labor safety.

7.4.4 Digital Content Regulation

In order to protect the rights of communication through information networks of copyright owners, performers, and authors of audio and video recordings, the right of communication through information networks was formally included in the revised "Copyright Law of the People's Republic of China" in October 2001; in May 2006, the "Regulations on the Protection of the Right of Communication Through Information Networks" was officially promulgated. In June 2014, the National Copyright Administration, together with the CAC, the MIIT, and the MPS, launched the "Sword Net 2014" special campaign, continuously carrying out special rectification of online infringement and piracy, cracking a number of key online infringement and piracy cases, and closing 200 websites suspected of infringement and piracy in accordance with the law.

China vigorously combats illegal activities such as telecommunications and online fraud and the infringement of citizens personal information. It has launched the "Operation Clean Net" special campaign. The "Operation Qinglang" special campaign has been implemented in depth, effectively curbing the spread of Internet chaos. From 2019 to 2022, China had cleared 20.49 billion pieces of illegal and harmful information and 1.39 billion accounts, removed more than 67,000 illegal and irregular applications, and closed more than 42,000 illegal websites through nearly 30 special campaigns. China continues to strengthen the construction of online information content, actively create a good network ecosystem, and promote the formation of a clean and sound cyberspace.

7.4.5 Information Technology Governance

China has formulated specialized legal regulations for the management of AI algorithms. In August 2021, China issued the "Guidelines for the Construction of National

New-Generation Artificial Intelligence Standard System" to regulate AI services and meet the requirements of traditional moral ethics and legal order. On August 27, 2021, the CAC issued a notice on soliciting public opinions on the "Regulations on the Management of Algorithmic Recommendations for Internet Information Services (Draft for Comments)", which specifically regulates the management of algorithm recommendation technology. In September 2021, the CAC and eight other ministries and commissions jointly issued the "Guiding Opinions on Strengthening the Comprehensive Governance of Internet Information Services Algorithms", which clearly requires enterprises using algorithms to bear the main responsibility for the results of algorithm application, establish an algorithm security responsibility system and a scientific and technological ethics review system, improve the algorithm security management organization, strengthen risk prevention and control and hidden danger investigation and governance, and improve the capability and level of responding to algorithm security emergencies.

China has also established regulations related to the security of Internet of Vehicles (IoV) and automotive data. In August 2021, the CAC and other departments jointly issued the "Several Provisions on the Administration of Automobile Data Security (Trial)", advocating the adherence to the principles of "in-vehicle processing", "non-collection by default", "applicable accuracy range", and "desensitization processing" to reduce the disorderly collection and illegal abuse of automobile data. On September 13, 2021, the MIIT issued the "Notice of the Ministry of Industry and Information Technology on Strengthening the Real-Name Registration Management of Internet of Vehicles SIM Cards", clarifying the requirements for real-name registration of IoV subscriber identity module (SIM) cards at different sales stages to prevent security risks in the IoV. On September 15, 2021, the MIIT issued the "Notice of the Ministry of Industry and Information Technology on Strengthening the Cybersecurity and Data Security of the Internet of Vehicles", which put forward basic requirements for the cybersecurity and data security of the IoV.

7.4.6 Digital Economy Regulation

In the early stages of the development of the platform economy of the digital economy, China adopted the development concept of "inclusiveness and prudence" and did not intervene too much or too early in corporate behavior. However, as platform enterprises reach a considerable scale and the industry becoming increasingly mature, strengthening supervision has become a necessary measure to maintain market competition and promote the survival of the fittest.

China has been actively building a digital development governance system to create a healthy and orderly development environment. It has formulated and implemented the "Electronic Commerce Law of the People's Republic of China", the "Anti-Unfair Competition Law of the People's Republic of China", the "Anti-Monopoly Law of the People's Republic of China", and the "Regulations on Optimizing the Business Environment". It has issued the "Guiding Opinions of the General Office of the State Council on Promoting the Standardized and Healthy Development of the Platform Economy" to promote the standardized and healthy development of the platform economy, as well as effectively safeguard the legitimate rights and interests of market entities and the people. In February 2021, the Anti-Monopoly Committee of the State Council issued the "Anti-Monopoly Guidelines of the Anti-Monopoly Committee of the State Council on the Platform Economy", clarifying the principles of anti-monopoly law enforcement in the field of platform economy. In March 2021, the SAMR issued the "Measures on the Supervision and Administration of Online Transactions" to further standardize online transaction activities, maintain online transaction order, and protect the legitimate rights and interests of all parties involved in online transactions.

China has formulated relevant regulations for new online marketing methods to keep pace with the times. In the field of online live streaming and marketing, at the end of 2020, relevant normative documents such as the "Guiding Opinions of the State Administration for Market Regulation on Strengthening the Supervision of Online Live Streaming Marketing Activities" and the "Notice of the State Administration of Radio and Television on Strengthening the Management of Online Show Live Streaming and E-commerce Live Streaming" were successively issued. On April 16, 2021, the CAC, the MPS, the Ministry of Commerce (MOFCOM), and other four ministries and commissions jointly issued the "Measures on the Management of Online Live Streaming Marketing (Trial)", which systematically and comprehensively solved the problems of e-commerce online live streaming marketing by issuing normative documents.

7.4.7 Collaborative Governance Model

The Chinese government has played a coordinating and leading role, actively involving various parties in the Internet governance. By employing a comprehensive approach that includes legal regulations, administrative management, industrial policies, technical standards, and public education, the government has mobilized efforts from all sides to collectively promote the Internet development. This has led to the establishment of an Internet governance model that not only aligns with global Internet norms but also embodies distinct characteristics tailored to China.

Internet enterprises conscientiously fulfill their principal responsibilities. In November 2004, Sina, Sohu, and NetEase announced the self-discipline details of the "Integrity Self-discipline Alliance" of the Chinese Internet industry, marking that China's Internet information service enterprises have taken the initiative to assume the principal responsibility for maintaining market order. In June 2013, 21 Internet enterprises including Alibaba, Tencent, Baidu, Sina, Shengda, and NetEase established the "Internet Anti-Fraud Committee" to promote the joint fight against online fraud across the network and jointly build a transaction security ecosystem. In 2013, search engine service enterprises like Baidu and Qihoo 360 successively launched plans to protect the rights and interests of Internet users, taking on greater social responsibilities.

Industry organizations have been effectively promoting the healthy development of the Internet. China attaches great importance to the important role of social organizations in the field of Internet governance, strongly supports the construction of various industry associations and social organizations, and actively participates in Internet governance. In May 2001, the Internet Society of China, a national industry organization, was established. It formulated and issued a series of self-discipline norms, including the "Self-Discipline Convention of China's Internet Industry", "Self-Discipline Norms for Internet Sites to Prohibit the Spread of Obscene and Pornographic and Other Harmful Information", "Self-Discipline Convention for Resisting Malware", and "Copyright Self-Discipline Declaration of China's Internet Industry". In November 2002, the Internet Society of China established the Anti-Spam Coordination Group, which made outstanding contributions to the governance of spam.

7.5 Summary

The Internet and the cyberspace derived from it are the remarkable achievements of the development of human civilization. In cyberspace, all kinds of activities of individuals and groups need to be protected and supervised within the law framework of human society. The concept and scope of cyberspace governance are constantly evolving and developing. Governments around the world shoulder the important mission of protecting the legitimate rights and interests of individuals and groups in cyberspace and maintaining the normal order of cyberspace.

China has established a complete legal system for cyberspace. From the initial connection to the Internet to becoming a world-leading power in mobile Internet application innovation, China's cyber legislation has achieved a historic leap from no-existence to gradually being sound, and from a single norm to comprehensive coverage. By June 2024, China had established more than 150 pieces of legislation governing cyberspace. These legislation cover many key areas, such as the protection of personal digital rights and interests, the improvement of digital economy rules, the maintenance of network information security, and the optimization of the cyber-ecological norms. China's legal system for cyberspace not only ensures the orderly development of China's cyberspace, but also provides rich reference examples for global cyber legislation.

China has built a strong force for protecting cybersecurity. China has been committed to the independent research and development of a series of cybersecurity technologies, overcoming many key security technical problems, developing a series of advanced security technologies and algorithms, and constructing a complete security technology standard system. China has established professional protection institutions and high-quality teams covering key security fields. With the ability of regular network monitoring and efficient emergency response, China can respond to various cybersecurity threats timely and properly. China's cybersecurity industry is showing a good momentum of vigorous development, the industry scale continues to grow steadily, and products cover many key fields, forming a complete and mature industrial chain. China's cybersecurity technologies and industry not only ensure the healthy operation of domestic networks and information systems, but also can provide professional security services for other countries.

China has carried out all-round and multi-dimensional cyberspace governance practices. In the field of public environment governance, China continuously improves the Internet legal environment, strengthens the management of Internet infrastructure resources, and strictly standardizes user account management, maintaining a well-ordered cyberspace. In terms of personal rights and interests protection, the data collection process has been standardized, and the personal privacy protection has been strengthened, ensuring that groups such as teenagers and the elderly can access the Internet equally and conveniently, and effectively safeguarding the legitimate rights and interests of workers in the digital economy. In addition, China continues to make efforts in areas such as digital content supervision, information technology governance, and digital economy regulation. The Chinese government plays a leading and coordinating role, mobilizing all parties related to the Internet industry to participate in cyberspace governance, forming a collaborative, efficient, and sustainable development governance model.

The booming development of the Internet has brought rare opportunities as well as many challenges to the world. The rule of law is undoubtedly the core element in Internet governance. Since connecting to the international Internet, China has comprehensively promoted the rule of law in cyberspace, established a sound legal system, and effectively promoted the positive interaction between cybersecurity and development. The path of governing the Internet according to law taken by China, has not only significantly improved the domestic Internet governance level, but also provided a comprehensive, systematic, and reference-worthy governance model for other countries in the world, contributing Chinese wisdom to promoting the improvement and development of the global Internet governance system.

Chapter 8
International Digital Cooperation

8.1 Demands and Challenges

Since the advent of the twenty-first century, the advancement of information and communications technology (ICT) has not only expanded the living space for humanity, but has also transformed the modes of societal production. However, it has not fundamentally altered the international political order. Various issues persist within the global governance of cyberspace, such as the disparities between the East and the West as well as the North–South divide. This urgently calls for nations across the globe to unite and collectively address these challenges.

8.1.1 Digital Divide in Internet Era

The Internet serves as the shared habitat of humanity, and its evolution reflects the level of a country's economic and technological development. Undoubtedly, the development of the global Internet exhibits disparities among nations and regions. This discrepancy is commonly referred to as the "digital divide".

The global digital divide persists and is widening. According to the data from the International Telecommunication Union, by the end of 2021, the number of global Internet users had reached 4.9 billion, approximately 63% of the global population, with an annual growth rate of 13.3% among Internet users in developing countries. Over the period from 2015 to 2021, the coverage of 4G networks doubled, encompassing 88% of the global population. In the Asia–Pacific region, Europe, and the Americas, 90% of the population can access to mobile broadband networks (3G or above). Between 2020 and 2021, Africa has witnessed a 21 percentage point increase in 4G network coverage. However, as of 2021, a significant gap still existed, with 18% of the population unable to access to mobile broadband networks, highlighting substantial disparities in coverage. Globally, in 2021, approximately 2.9

W. Liu et al. (eds.), *Digital China*, Introduction to China's S&T Innovation,
https://doi.org/10.1007/978-981-95-6240-4_8

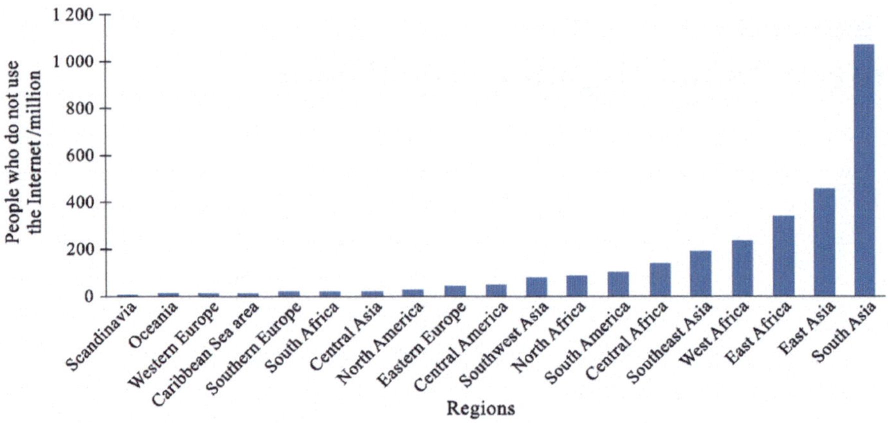

Fig. 8.1 Global distribution and number of people who do not use the Internet (2022). *Data Source* Digital 2022: Global Overview Report

billion people still lack access to the Internet, with 96% located in developing countries, and around 390 million individuals residing in areas devoid of mobile broadband signal coverage. The global distribution and number of people who do not use the Internet in 2022 are shown in Fig. 8.1.

Developed nations, leveraging their technological advantages, have secured the authority to formulate rules and control the discourse within the realm of the Internet. It is difficult for the developing countries in the digital sphere to participate in an equal footing in various global cyberspace governance affairs. They are technologically subordinate and economically subjected to constraints and exploitation. This highlights the presence of imbalances and inadequacies in the development of cyberspace, calling for the establishment of just and equitable rules and mechanisms on a global scale. This initiative aims to safeguard the right of developing countries to participate equally in global cyberspace affairs and achieve sustainable development in the digital realm.

8.1.2 Global Cybersecurity Risk

In recent years, the global cybersecurity landscape has become increasingly perilous, marked by numerous significant incidents of cybersecurity threats, such as distributed denial of service (DDoS) attacks, ransomware attacks, telecommunications fraud, and malicious code attacks on mobile terminals. According to the "Global Risks Report 2024" released by the World Economic Forum, cyber insecurity ranks among the top 10 global risks of most concern over the next decade. In 2021, cyberattacks frequently triggered disruptions in Internet services and halted factory operations, profoundly affecting social stability and the daily lives of citizens.

For instance, in May 2021, Colonial Pipeline, the largest fuel pipeline operator in the United States (US), fell victim to a Darkside ransomware attack, severely impacting fuel supplies along the US East Coast. In July 2021, the automatic ticketing system of the Northern Railways in the United Kingdom (UK) experienced a ransomware attack, leading to the paralysis of the ticketing network. In September 2021, Vocus, a telecommunications enterprise in New Zealand, faced a DDoS attack, resulting in disruptions in network services in cities such as Auckland and Wellington.

Cyberterrorism proliferates unchecked in cyberspace, and traditional criminal activities are increasingly becoming networked. Consequently, many nations have elevated the importance of cybersecurity to the level of a national strategy. In 2014, the National Institute of Standards and Technology (NIST) in the US released the "Framework for Improving Critical Infrastructure Cybersecurity", defining cybersecurity as the "process of protecting information by preventing, detecting, and responding to attacks". In the same year, the Russian Federation, in its "Concept of the Russian Federation Cybersecurity Strategy", defined cybersecurity as the "sum of all the conditions to prevent potential threats and their consequences from affecting all components of cyberspace". In 2020, the European Union (EU) issued "The EU's Cybersecurity Strategy for the Digital Decade", encompassing relevant recommendations in terms of regulations, investments, and policy tools.

The continual iteration of new technologies and their widespread social applications pose novel governance obstacles and challenges for the international community. With the emergence and proliferation of technologies such as big data and artificial intelligence, cybersecurity faces unprecedented technical challenges. In November 2020, the World Economic Forum released a report titled "Future Series: Cybersecurity, Emerging Technologies and Systemic Risks". The report stated that with the development of emerging technologies, the scale, speed and connectivity of cyberspace are undergoing major changes, which brought about a series of new systemic risks and challenges. Therefore, it is necessary to establish cybersecurity norms that are generally recognized by all countries, put transnational cybercrime issues on the track of the rule of law, and safeguard the legitimate rights and interests of cyber actors in all countries.

8.1.3 Global Digital Trade Demands

The new generation of ICT is accelerating its extension into the international economic and trade sectors, giving rise to digital trade, which has profound impacts on trade methods, trade objects, trade rules, and trade landscape. From 2011 to 2020, global digital trade experienced rapid development, with a compound annual growth rate of 4.4% in digital services trade, significantly surpassing that of traditional international services trade (1.19%) and goods trade (—0.4%). The proportion of digital trade in global trade continues to rise. Since 2012, over 90% of services trade agreements have started to incorporate relevant clauses or dedicated chapters on digital trade (e-commerce). Simultaneously, international digital trade encounters numerous challenges. For example, the digitalization of trade objects and methods has led to the emergence of many new rule vacuums that urgently need to be addressed and perfected, and the existing trade rules and negotiation mechanisms also require modification and updates.

With the evolution of ICT, an increasing amount of goods trade is transforming into digitized, virtual services trade. While digital technology can benefit small and medium enterprises (SMEs) in developing countries by reducing trade costs and lowering entry barriers, these nations face greater challenges in engaging in digital trade due to the influence of technological industry development trends and super platforms. For instance, when SMEs in developing countries conduct international trade through online platforms, they may encounter unfair treatment from online platform services providers. Shortcomings in digital infrastructure and data management experience of developing countries can also result in losses. For example, due to a lack of appropriate technological regulatory measures, developing countries are at risk of erosion in tariff revenue from digital trade.

8.2 Community with a Shared Future in Cyberspace

The "community with a shared future in cyberspace" is an extension of the concept of "community with a shared future for humanity" into the realm of the Internet. In March 2013, Chinese President Xi Jinping first introduced the significant concept of "community with a shared future for humanity" during his visit to Russia. On November 23, 2020, the World Internet Conference·Internet Development Forum opened in Wuzhen City, Zhejiang Province. Chinese President Xi Jinping sent a congratulatory letter to the forum, in which he pointed out that China is willing to work with all countries in the world to seize the historical opportunities of the information revolution, cultivate new impetus for innovation and development, create a new situation for digital cooperation, foster a new landscape of cybersecurity, build a community with a shared future in cyberspace, and work together to create a better future for mankind.

8.2.1 China's Voice in Cyberspace Governance

Since 2014, China has consecutively hosted the World Internet Conference (WIC) in Wuzhen City, Zhejiang Province for 11 years. This conference serves as both establishing an international platform for connectivity and cooperation between China and the world in the field of the Internet and a Chinese platform for shared governance and collaboration in cyberspace. Representatives from various sectors, including governments, international organizations, Internet enterprises, think tanks, industry associations, and technical communities, have been invited to participate in the conference for exchanges.

The organizing committee of the conference has successively released a concept document named "Jointly Build a Community with a Shared Future in Cyberspace", as well as the "Initiative on Jointly Building a Community with a Shared Future in Cyberspace". It has also organized case presentations and demonstrations to elaborate on the implementation of the concept of constructing a community with a shared future in cyberspace. The successful hosting of the conference has significantly promoted close connections and in-depth communication in the global Internet, effectively advancing China's experience, solutions, and ideas in building a community with a shared future in cyberspace from conceptual consensus to concrete practices, and propelling the global Internet governance system toward a more just and equitable direction.

In August 2021, the China-Africa Internet Development and Cooperation Forum, themed "Seeking Development, Ensuring Security, and Building a Community with a Shared Future in Cyberspace", was conducted through video linkages. Representatives from 14 African countries and the African Union Commission attended the forum, engaging in extensive discussions on topics such as sharing the dividends of digital technology and jointly safeguarding cybersecurity. China initiated the "China-Africa Initiative for Building a Community with a Shared Future in Cyberspace", calling for the participation of governments, Internet enterprises, technical communities, social organizations, and individual citizens from both China and African countries to collaboratively build a more closely-knit community with a shared future in cyberspace.

China has been continuously promoting international exchanges and cooperation in cyberspace. By leveraging the United Nations (UN) as the primary channel and adhering to the United Nations Charter (UN Charter), China actively participates in the formulation of international rules in the digital and cyberspace realms. China also provides digital assistance, including technology, equipment, and services to less developed countries, enabling them to share the benefits of the digital age and supporting the realization of The 2030 Agenda for Sustainable Development.

8.2.2 Building a Community with a Shared Future in Cyberspace

China's advocacy for a community with a shared future in cyberspace is rich in connotations. China adheres to the global governance principle of achieving shared growth through consultation and collaboration, promoting the establishment of a multilateral, democratic, and transparent system of the international cyberspace governance. China proposes that the future international cyberspace should be characterized by innovation, security, order, equality, respect, openness, and sharing. All countries worldwide are encouraged to jointly promote development, collectively safeguard security, participate in governance, and share the achievements in cyberspace. The concept of a community with a shared future in cyberspace can be interpreted as a community in terms of development, security, responsibility, and interests.

Building a community of development. Addressing the imbalances in the development levels of different countries and regions in aspects such as Internet penetration, infrastructure construction, technological innovation, digital economic development, and digital literacy and skills is crucial, which affects and limits the informatization efforts and digital transformation of countries worldwide, particularly developing countries. The emphasis on building a community of development involves adopting more proactive, inclusive, coordinated, and beneficial policies. This aims to accelerate the global popularization of information infrastructure and provide affordable and quality network services for developing countries. It involves maximizing the role of the digital economy as an engine for global economic development and actively promoting the development of digital industries and the transformation of industrial digitalization.

Building a community of security. Security is considered a prerequisite for development, and a secure, stable, and prosperous cyberspace is of significant importance to all countries globally. Given that cybersecurity is a challenge of global scale, no country can remain unaffected or isolated. Building a community of security involves advocating an open and cooperative cybersecurity ideology, emphasizing the dual importance of security and development, and encouraging both encouragement and regulation. It entails strengthening the protection of critical information infrastructure and promoting international cooperation on data security. It also involves maintaining the neutrality of ICT and globalizing industrial security to jointly curb the misuse of ICT. Establishing strategic mutual trust, timely sharing information on cybersecurity threats, effectively coordinating major cybersecurity incidents, cooperating in combating cyberterrorism and cybercrime, and collectively safeguarding the peace and security of cyberspace are integral to this community.

Building a community of responsibility. Cyberspace is a common space for human activities, and the future and destiny of cyberspace should be jointly controlled by all countries in the world. Building a community of responsibility means adhering to multilateral participation and multi-party participation, and actively promoting the reform and construction of the global Internet governance system. It entails giving full play to the UN as the main channel for international governance in cyberspace, and utilizing the roles of various subjects such as governments, international organizations, Internet enterprises, technology communities, social organizations, and individual citizens to establish cooperation based on mutual trust and coordinated order. It also involves improving the dialogue and consultation mechanism, jointly studying and formulating cyberspace governance norms, reflecting the interests and concerns of all parties in a more balance manner, especially the interests of the vast number of developing countries, and making the governance system more fair and reasonable.

Building a community of interests. The achievements of cyberspace development and governance should be shared globally to ensure that different countries, different ethnic groups, and different populations have equal access to the benefits of cyberspace development. Building a community of interests involves facilitating the use of new-generation ICT by SMEs to promote innovation in products, services, processes, organizations, and business models. This ensures that SMEs can share more opportunities from digital economic development. The community emphasizes the protection of vulnerable groups in cyberspace, strengthens the development of cyber ethics, and fosters a civilized cyberspace. Enhancing the network development capabilities of developing countries, bridging the digital divide, and promoting a more inclusive global development are integral to this community.

8.3 International Cooperation on Digital Economy

China actively engages in international cooperation in the digital economy. It participates in the construction of Internet information infrastructure in various regions, promotes the internationalization of various new ICT and their applications, and contributes to the improvement of global digital connectivity. China also collaborates with various countries in the field of digital trade, fostering the integrated development of the global digital economy and the real economy.

8.3.1 Global ICT Infrastructure Development

8.3.1.1 Management of Internet Infrastructure Resources

The Internet infrastructure resources encompass IP addresses, domain names, and more. China places great emphasis on communications and collaboration with international Internet organizations, such as the Internet Corporation for Assigned Names and Numbers (ICANN), Asia–Pacific Network Information Centre (APNIC), World Wide Web Consortium (W3C), Internet Society (ISOC), Internet Architecture Board (IAB), and Internet Engineering Task Force (IETF). This effort aims to promote the rational allocation of critical Internet resources and the interconnection of basic Internet services, such as domain name service (DNS).

In 2002, the ICANN meeting was held in Shanghai, China for the first time. In 2003, Qian Hualin, a researcher from the Chinese Academy of Sciences, was elected as an ICANN director. In November 2020, Yao Jiankang, a technical expert from the China Internet Network Information Center (CNNIC), was elected as a new director of the Country Code Names Supporting Organization (ccNSO), playing an active role in the organization's relevant policy formulation, agenda setting, and international exchanges and cooperation, and enhancing China's activity and recognition in the international technology community.

The Asia Pacific Top level Domain Association (APTLD) is an organization of country code top-level domains (ccTLDs) in the Asia–Pacific region. Its purpose is to promote information exchange on the development of Internet domain name application technology among its members, and enhance the influence of these members in the global Internet field. As of 2021, APTLD had 50 formal institutional members and 19 associate members. In February 2021, a CNNIC representative was elected as the member of the new APTLD board, making important contributions to further maintaining international community relations and deepening regional exchanges and cooperation.

APNIC is one of the five major regional Internet IP address registration and management institutions worldwide, responsible for the allocation and management of IP addresses in the Asia–Pacific region. All of China's IP address resources are obtained through APNIC. In March 2021, a CNNIC representative was elected as the member of the APNIC Executive Committee. China took this opportunity to deepen cooperation and exchanges with APNIC and promote cooperation with the countries along the Belt and Road.

8.3.1.2 African ICT Infrastructure Development

Over the years, China has been actively engaging in infrastructure development in Africa, aiding the implementation of numerous projects for both basic infrastructure and public welfare. From the establishment of the Forum on China-Africa Cooperation (FOCAC) in 2000 to 2021, China had added and upgraded more than 10,000 km of railway networks, nearly 100,000 km of roads, 120 million kilowatts of installed power capacity, and 150,000 km of communications backbone networks in Africa. As of June 2019, about 80% of the backbone networks in Africa were invested by Chinese funded enterprises; among the 54 African countries, more than 40 countries have adopted Chinese products and technologies in communications network facilities.

In 2018, the Beijing Summit of the FOCAC was held in Beijing, and it released the "Forum on China-Africa Cooperation Beijing Action Plan (2019–2021)". China and African countries encourage and support their respective enterprises to cooperate in the construction of communications infrastructure such as optical cable backbone networks, cross-border interconnection, international submarine optical cables, new-generation mobile communications networks, and data centers in African countries. The two sides will actively explore and promote the application of new technologies such as cloud computing, big data, and mobile Internet, and enhance the role of ICT in maintaining social order, counter-terrorism, and combating crime.

With the participation of Chinese enterprises, the level of broadband network infrastructure in some African countries has been significantly improved. For example, Chinese enterprises had undertaken the construction of the Glo-2 submarine optical cable system of Nigerian telecom operator Globacom. The system will be laid along the coast of Nigeria, connecting Alpha Beach in Lagos to the southern region, with a total length of 850 km and providing a transmission capacity of 12 Tbps.

With the help of Chinese enterprises, Africa is also accelerating into the 5G commercial era. In July 2020, South African telecom operator Rain successfully launched Africa's first 5G commercial network with the technical support of Huawei.

In November 2022, Zambian mobile communications operator MTN launched the country's first 5G commercial network services with the support of Huawei. In April 2023, Chinese enterprises and partners such as Ugandan telecom operator MTN and HIMA Cement Company jointly launched the Ugandan's first 5G digital cement factory project supported by the Chinese-funded enterprise in Kampala, the capital of Uganda, to achieve real-time data transmission in the production area.

8.3.2 Provision of Global Digital Services

Besides the network infrastructure, China also actively participates in the digital industrialization processes of various countries, providing globally-oriented information technology services to nations worldwide. This includes services like the BeiDou Navigation Satellite System (BDS) and cloud services.

BDS has become an important global space–time infrastructure. China continues to carry out satellite navigation cooperation and exchanges with regions, countries, and regional organizations such as the League of Arab States (LAS), the Association of Southeast Asian Nations (ASEAN), Central Asia, and Africa. Beidou-related products have been exported to more than half of the countries and regions in the world. In 2021, at the first China-Africa BDS Cooperation Forum held in Beijing, the "10 Application Scenarios of BDS in Africa" was officially released, showing BDS's application cases and application solutions in the fields of road transportation, vehicle management, railway industry, precision agriculture, etc. in Africa.

The China-Arab BDS/GNSS Center was located in Tunis, the capital of Tunisia. It was China's first overseas BDS center, providing satellite navigation training, test evaluation, and technical research for African and Arab countries. The center's central screen can display the collected data in real time, and Tunisia can receive signals from more than 12 BeiDou satellites on average. In Mozambique, the China-Africa Saisai Agricultural Cooperation Project widely used drones with BDS positioning for field plant protection operations such as farmland mapping, rice sowing, and pesticide spraying. Compared with traditional manual spraying of only 3 to 4 mu per hour, plant protection drones can spray pesticides on hundreds of acres per hour, and the cumulative operation has exceeded 30,000 acres.

In addition, China's cloud computing platform has also begun to provide international services, especially for Africa, the Middle East, Southeast Asian countries and participating countries in the Belt and Road Initiative (BRI). The "Global Microbial Resource Data Sharing Platform", a global microbial big data platform infrastructure led by China, had gathered 520,000 microbial physical resource data from 141 partners in 51 countries as of 2021, forming an interconnected cooperation network, establishing a global microbial strain collection catalog, and promoting the effective use of global microbial data resources.

8.3.3 Cooperation on Global Digital Economy

China actively engages in international and regional multilateral mechanisms for cooperation on digital economic governance, advocating and advancing various initiatives and declarations. It puts forth proposals aligning with the interests and demands of the majority of countries worldwide, strengthened collaboration with specialized international organizations, and contributed to global digital economic governance.

Promoting the digital economy cooperation process of the Asia–Pacific Economic Cooperation (APEC). In 2014, as the host of APEC, China introduced the Internet economy into the APEC cooperation framework for the first time, initiating and promoting the adoption of the "APEC Initiative of Cooperation to Promote the Internet Economy". In 2019, after the establishment of the APEC Digital Economy Steering Group, China actively promoted the comprehensive and balanced implementation of the "APEC Internet and Digital Economy Roadmap". Since 2020, China has successively put forward initiatives such as "optimizing the digital business environment and activating the vitality of market entities", all of which have been unanimously adopted by APEC.

Actively participating in digital economy cooperation under the G20 framework. In 2016, the 11th G20 Summit was held in China. Driven by China, the meeting listed "digital economy" as an important topic in the G20 Innovation Growth Blueprint for the first time, and adopted the "G20 Digital Economy Development and Cooperation Initiative", which is the world's first digital economy policy document signed by leaders of many countries. Since then, digital economy has become one of the core topics of the G20. In recent years, China has actively participated in the G20 Digital Economy Ministerial Meeting and relevant consultations of the Digital Economy Task Force (DETF), and promoted the upgrade of the DETF to a working group to ensure that the achievements of digital economic development benefit people around the world.

Continuously expanding exchanges and cooperation in the digital economy among BRICS countries. In 2017, the ninth BRICS Leaders Meeting was held in China. The "BRICS Leaders Xiamen Declaration" adopted at the meeting clearly proposed to deepen pragmatic cooperation in the fields of information and communications technology, e-commerce, and Internet space. In 2019, the China Branch of BRICS Institute of Future Networks was officially unveiled in Shenzhen. In 2022, the 14th BRICS Leaders Meeting adopted the BRICS Digital Economy Partnership Framework. In addition, China held important events such as the BRICS Digital Economy Dialogue, opening a new process of BRICS digital economy cooperation.

Deepening digital economy cooperation with ASEAN. In 2020, the China-ASEAN Year of Digital Economy Cooperation was held, with the theme of "Combating COVID-19 through Joint Efforts,Embracing Cooperation through ICT and Digital Development" and a network affairs dialogue. The 23rd China-ASEAN Leaders Meeting issued the "China-ASEAN Initiative on Establishing a Digital

Economy Cooperation Partnership", and agreed to further deepen cooperation in the field of digital economy.

Actively promoting digital economy cooperation in the World Trade Organization (WTO). In 2017, China officially announced its accession to the WTO's "Friends of E-commerce Development" and worked with developing members to support the WTO's e-commerce consultations. In 2019, China and other 75 WTO members including the US, the EU, Russia, Brazil, Singapore, Nigeria, and Myanmar, jointly issued a "Joint Statement on E-commerce" to launch negotiations on trade-related e-commerce issues. In 2022, China and other WTO members jointly issued the Ministerial Decision on the E-commerce Work Plan, supporting the exemption of tariffs on electronic transmissions and promoting the development of the global digital economy.

8.3.4 International Cross-Border E-commerce

China is a major exporter of cross-border e-commerce and the largest import market for cross-border e-commerce in the world. With the vigorous development of cross-border e-commerce, China is actively strengthening coordination and interaction with other countries in the field of international laws and regulations for cross-border e-commerce, and deepening cooperation with other developing countries and economically backward countries in the field of e-commerce, helping SMEs in various countries to narrow the digital divide and form a win–win development pattern of international cooperation. The development of China's cross-border e-commerce has developed from buying and selling globally to the stage of international ecological cooperation. The development of cross-border e-commerce not only sells products to end consumers in overseas markets, but also integrates various domestic and foreign resources, integrates market entities in different countries, forms a cross-border e-commerce ecological service system that transcends national boundaries, and promotes the development of global digital trade.

In October 2021, the Ministry of Commerce, the Office of the Central Cyberspace Affairs Commission, and the National Development and Reform Commission jointly issued the "14th Five-Year Plan for E-commerce Development", and put forward development goals in the relevant part of cross-border e-commerce, namely, to increase the transaction value of cross-border e-commerce from 1.69 trillion yuan in 2020 to 2.5 trillion yuan by 2025. In 2023, China's cross-border e-commerce imports and exports totaled 2.38 trillion yuan, with an annual growth rate of 15.6%; specially, exports were 1.83 trillion yuan, with an annual growth rate of 19.6%; imports were 548.3 billion yuan, with an annual growth rate of 3.9%. The number of Chinese consumers participating in cross-border e-commerce imports has been increasing year by year, reaching 163 million in 2023.

8.3.4.1 Policies for Cross-Border E-commerce Cooperation

China actively promotes international cooperation in cross-border e-commerce, participates in e-commerce consultations under multilateral and regional trade mechanisms such as the WTO, the G20, the APEC, the BRICS, and the Shanghai Cooperation Organization (SCO), jointly builds regional high-level digital economy rules with free trade partners, and has made breakthroughs in the construction of international e-commerce rules. The e-commerce chapter in the "Regional Comprehensive Economic Partnership (RCEP)" has become the international e-commerce rule with extensive coverage area, comprehensive content, and a high level of expertise. In 2022, the top 10 countries in terms of China's cross-border e-commerce exports are the US, Malaysia, Singapore, Australia, Vietnam, South Korea, Thailand, the Philippines, India, and Japan; the top 10 countries (regions) in terms of cross-border e-commerce imports are China's Hong Kong, South Korea, Japan, the US, Australia, the Netherlands, New Zealand, Germany, France, and the UK.

In conjunction with the BRI, China has proposed a bilateral international cooperation mechanism for "Silk Road E-commerce". By signing a bilateral e-commerce cooperation memorandum, China has formulated a cross-border e-commerce strategy based on the economic development characteristics of the partner countries, giving play to China's advantages in e-commerce technology application, model innovation, and market scale. On November 4, 2022, Chinese President Xi Jinping proposed in his speech at the opening ceremony of the fifth China International Import Expo (CIIE) to create a "Silk Road E-commerce" cooperation pilot zone and promote high-quality Belt and Road cooperation. As of September 2023, China had signed bilateral e-commerce cooperation memoranda with 30 countries. "Silk Road E-commerce" has become a new channel and new highlight of international economic and trade cooperation, and its partner countries are spread across the five continents of the world. In 2023, the cross-border e-commerce import and export value between China and the "Silk Road E-commerce" partner countries accounted for more than 30% of China's total cross-border e-commerce value.

The cooperation mechanism of "Silk Road E-commerce" is conducive to countries sharing the opportunities presented by the Chinese market. At previous CIIEs, the "Silk Road E-commerce" partner countries used the CIIE as a window to set up national pavilions to display their most distinctive and high-tech products, and quickly launch cross-border e-commerce platforms, obtaining huge Chinese orders. Hongqiao Pinhui is the Shanghai import commodity exhibition and trading center and the permanent exhibition center of the CIIE. In 2022, Hongqiao Pinhui built two important sub-platforms, Shanghai International Friendship City Port and Live E-commerce Base, attracting nearly 5000 products from 36 partner countries. In 2023, Hongqiao Pinhui opened the "Silk Road E-commerce" online country pavilion to create convenience for high-quality enterprises and products from partner countries to enter the Chinese market. In the first half of 2023 alone, 42 suppliers from partner countries were introduced.

8.3.4.2 Measures for Cross-Border E-commerce Development

The management of cross-border e-commerce involves the management of multiple processes such as transactions, payments, logistics, customs clearance, tax refunds, and foreign exchange settlement. In order to promote the development of the cross-border e-commerce industry, China is actively exploring a more efficient cross-border e-commerce management mechanism. China has established cross-border e-commerce comprehensive pilot zones in many cities and regions, aiming to promote the pioneering trials of cross-border e-commerce technical standards, business processes, regulatory models, and informatization construction. Starting from the State Council's approvement to establish the China (Hangzhou) Cross-border E-commerce Comprehensive Pilot Zone in March 2015, as of November 2022, the number of China's cross-border e-commerce comprehensive pilot zones had reached 165, covering 31 provinces, autonomous regions, and municipalities. Taking the China (Hangzhou) Cross-border E-commerce Comprehensive Pilot Zone as an example, the region has carried out cross-border e-commerce "small package export", "direct mail import", "online shopping bonded import", as well as cross-border business-to-business (B2B) export, bonded export and other business pilots, and took the lead in exploring cross-border e-commerce return and exchange centers, "global central warehouses", fixed-point distribution, "bonded import + retail processing" and other initiatives. As of June 2023, the scale of cross-border e-commerce import and export in China (Hangzhou) Cross-border E-commerce Comprehensive Pilot Zone had expanded from 120 million yuan in the early stage of its establishment to more than 120 billion yuan, and the number of cross-border e-commerce enterprises had jumped from more than 200 to more than 55,000, gathering 2/3 of the country's cross-border e-commerce platforms. Additionally, cross-border payment institutions in Hangzhou had served 1.5 million sellers nationwide, and their transaction value accounting for 70% of the country's cross-border payment transaction value.

The implementation of cross-border e-commerce is inseparable from the construction of a cross-border international logistics system. "China Railway Express (CR Express)" and "Silk Road Maritime (SRM)" are important platforms for China to promote international comprehensive logistics services. CR Express is an international railway combined transport train for containers between China and Europe and countries along the Belt and Road. In March 2011, the first CR Express departed from Chongqing, China to Duisburg, Germany. After more than 10 years of development, CR Express has opened more than 70 operating routes, which can reach more than 160 cities in 22 European countries, becoming the backbone of international logistics land transportation. As of November 2023, a total of 77,000 CR Express train trips had been launched, with a total mileage of more than 700 million kilometers. SRM is an international shipping integrated service platform initiated by China to connect global ports and shipping, which can realize data sharing and information exchange among ports, shipping, logistics, trade, and other enterprises. In December 2018, the first container route named SRM was launched from Xiamen, China. By April 2024, there were already 122 SRM named routes, connecting 135 ports in 46 countries and regions around the world, including 106 ports in 37 countries and regions that jointly build the Belt and Road. Through sea-rail transport, the function of ports can be extended to inland cargo sources. Cross-border e-commerce goods in Jiangxi Province, Hunan Province, and other places can be directly shipped to the sea after arriving at the port through the "one container to the end" model, shortening the transportation time. Chinese enterprises are also accelerating the construction of overseas service networks, especially overseas warehouses. The so-called "overseas warehouse" refers to domestic enterprises transporting goods to the target country by sea, land, air and other means of transportation, building their own warehouses locally or cooperating with platform warehouses and third-party overseas warehouses to sort and ship goods in local warehouses according to sales orders, which is much more efficient than direct domestic delivery. As of May 2024, Chinese enterprises have built more than 2500 overseas warehouses with an area of more than 30 million square meters, of which more than 1800 overseas warehouses focus on serving cross-border e-commerce, with an area of more than 22 million square meters.

Relying on cross-border e-commerce comprehensive pilot zones, cross-border logistics platforms and other industrial promotion measures, various places of China have focused on developing characteristic cross-border e-commerce industries. Taking Xi'an City, Shaanxi Province, as an example, this city is located in inland China and is an important stop for CR Express in the northwest region. Xi'an International Port Station covers an area of about 500 football fields, with an annual designed container throughput of 5.4 million twenty-foot equivalent units (TEUs) and a railway freight volume of 66 million tons. It operates 18 international trunk lines from Xi'an to Almaty in Kazakhstan, Tashkent in Uzbekistan, Hamburg in Germany, as well as lines crossing the Caspian Sea, covering the entire Eurasian continent. On average, a CR Express departs or arrives from Xi'an every 100 min. In 2015, Xi'an International Trade and Logistics Park was approved as a national e-commerce demonstration base, it vigorously developed the cross-border e-commerce industry, and approved for 15 national demonstration projects such as the cross-border e-commerce comprehensive pilot zone and the import trade promotion innovation demonstration zone. Additionally, it gradually built a complete cross-border e-commerce industry ecosystem led by various e-commerce leading enterprises such as Alibaba, JD. com, Amazon, eBay, and Douyin. By the end of 2023, Xi'an International Trade and Logistics Park had gathered 322 cross-border e-commerce and upstream and downstream enterprises, with the annual cross-border e-commerce transaction value exceeding 4 billion yuan.

8.3.4.3 Cross-Border E-commerce Exports

China's manufacturing industry ranks first in the world. In 2023, China exported 23.51 trillion yuan of manufacturing products, of which cross-border e-commerce transactions increased year by year. China's large e-commerce platforms have set up branches overseas. For example, JD.com has established an e-commerce platform for global consumers, JD Global Sales, through which merchants can sell products to more than 200 countries and regions. JD.com has established branches in the

US, Indonesia, Latin America, Europe and other countries and regions, and has also established multiple sub-websites in Russian, Spanish, English and other languages. Additionally, JD.com has set up more than 110 overseas logistics warehouses on five continents to reduce unnecessary logistics costs for global shopping. While the scale of cross-border e-commerce is expanding, a group of new-generation cross-border e-commerce platforms are developing rapidly in overseas markets, becoming a "Chinese force" that cannot be ignored. Today, these cross-border e-commerce platforms have a growing share of the global market and have become a new shopping choice for global consumers. Among the cross-border e-commerce platforms, Shein, Temu, AliExpress, and TikTok Shop are the most eye-catching. According to statistics from the Shanghai Pudong International Airport Customs, the export declaration volume of the three e-commerce platforms—Temu, TikTok, and Shein—in 2023 saw a more than tenfold increase compared to 2022.

Relying on the successful experience of Chinese e-commerce, the new cross-border e-commerce platforms have integrated China's domestic supply chain and quickly occupied the e-commerce market in many countries. In September 2022, Temu went online in the US and put forward the slogan "Shop like a billionaire". On the one hand, Temu replicated the marketing model employed by its parent company Pinduoduo in China to acquire users, and implemented it in the US market, gathering and gathered a large number of user orders; on the other hand, Temu launched "full trusteeship" and "flexible supply" services for China's SMEs. For merchants and enterprises settled in, Temu launched a support policy of zero commission and zero deposit, and provided warehousing, logistics, afte-sales and other infrastructure services. The special team of the "Duoduo Overseas Support Plan" has successively visited more than 100 small commodity industrial belts in more than 10 provinces in China, and organized the export of cost-effective goods overseas. Relevant data show that the prices of shoes, clothing, daily necessities and other supplies on Temu are usually 30% to 50% lower than those of its competitors. Duoduo Cross-border exports more than 400,000 parcels per day, with an average daily cargo weight of about 600 tons. As of December 2023, Temu's global unique visitors had reached 467 million, ranking second globally in cross-border e-commerce, after Amazon; about 18% of American households had shopped on Temu. In 2023, Temu brought Pinduoduo more than 25 billion yuan in revenue.

At the same time, Chinese entrepreneurs have expanded overseas markets. Founded in Kenya in 2014, Kilimall is an e-commerce platform that entered the African market early and has occupied and maintained a leading position. By referring to China's experience in developing e-commerce, the platform, based on the concept of taking root in Africa and coexisting and prospering with local African partners, has gradually built six major service systems, including Kilimall e-commerce transaction services, LipaPay payment services, KiliExpress logistics services, KiliShop community service stores, KiliWarehouse warehousing services, and KiliClick marketing services. Kilimall has a leading market share in Africa, especially in East Africa, where its market share remains above 50%. By the end of 2022, the platform had more than 10 million registered users, and more than 7.2 million annual active users, with an annual repurchase rate of more than 50%, making it an e-commerce platform with significant influence in Africa. Kilimall serves more than 8000 enterprises, has opened more than 12,000 stores with more than 1 million products, and has created more than 5000 jobs. It not only brings high-quality and low-priced Chinese products to the African people, but also promotes the online sales of local SMEs in Africa.

8.3.4.4 Cros-Border E-commerce Imports

China has a large middle-income group and is also one of the most promising markets in the world. In 2023, China imported more than 5 trillion yuan of bulk commodities, nearly 3 trillion yuan of electronic components and nearly 3 trillion yuan of consumer goods, providing a broad market space and cooperation opportunities for enterprises from all over the world. In 2023, China imported 1.95 trillion yuan of specialty foods, maternal and child products, digital home appliances and other consumer goods, an increase of 1.2%; among them, cross-border e-commerce imports was 548.3 billion yuan, representing an increase of 3.9%. The "White Paper on China's Cross-Border Import Consumption Trends in 2024" released by Nielsen IQ showed that the scale of China's cross-border import e-commerce increased from 444.1 billion yuan to 548.3 billion yuan from 2018 to 2023, and the number of users of China's cross-border import e-commerce increased at a compound annual growth rate of nearly 20% from 2017 to 2023, reaching 188 million people in 2023.

China's major e-commerce platforms such as Tmall and JD.com have opened international channels, specializing in cross-border e-commerce imports. Taking Tmall Global as an example, it was officially launched in February 2014. From the initial cooperation with 100 overseas brands, it has gradually grown into a cross-border e-commerce platform that includes more than 46,000 global brands from more than 90 countries and regions around the world, covering more than 7000 product categories, and serving more than 100 million consumer groups. Tmall Global took the lead in launching a bonded warehouse network and related infrastructure construction. Overseas goods can be temporarily stored in bonded areas and supervised by customs and inspection and quarantine departments. After an order is generated, the goods are shipped from the bonded area and can be delivered to consumers in one to three days. By the end of 2023, Tmall Global had built six major procurement centers in overseas regions, and its supply chain logistics network had covered more than 100 overseas warehouses, 500 international transportation trunk lines by sea, land and air, 40 core ports, and more than 100 bonded warehouses.80% of the goods shipped from Tmall Global bonded warehouses support door-to-door delivery services, covering more than 260 cities in China, and more than half of the orders in China can be delivered "next day"; among overseas direct mail products, more than half of the orders can be delivered within seven days.

The development of China's cross-border e-commerce has also driven the rapid development of related industries in the source countries. The most representative case is the Chilean cherry industry. Chile is located in the southern hemisphere, and its fruit supply period coincides with the winter market in China when fruits are scarce. In the past, the cost of transporting Chilean cherries to China was high, with challenges in the routes, storage, and preservation of bulk transportation. Driven by the demand for winter fruits in China, China's cross-border e-commerce and logistics platforms have actively adjusted and responded. China COSCO Shipping Corporation Limited and other freight enterprises have opened the "Cherry Express" connecting the Port of San Antonio in Chile and Shanghai in China, shortening the original 32 day voyage to 23 days, and increasing the supply of sea borne cherries by 1.8 times. Cross-border e-commerce platforms such as JD.com and Tmall Global have deployed direct procurement in Chile, monitoring the entire supply chain, and minimizing the time it takes for cherries to arrive from the place of production to consumers. Under various measures, Chilean cherries have gradually become a hot-selling winter fruit on Chinese e-commerce platforms. According to statistics from the Chilean Fruit Exporters Association, Chile's cherry exports to China grew by an average of 29% annually from 2016 to 2023, with 92% of cherries shipped to China from the end of November to February of the following year. Data released by the Chilean Agricultural Research and Policy Office show that the planted area of cherry cultivation in Chile has grown rapidly from 3241 hectares in 2000 to 67,570 hectares in 2023. The hot sales of cherries in the Chinese market have in turn promoted agricultural production in Chile.

The development of China's cross-border e-commerce has also explored new ways to eliminate poverty through cross-border trade. A typical case is Ethiopian coffee. In the past, Ethiopian coffee beans were mainly exported as raw beans in general trade. In addition, due to the restrictions of traditional trade rules, trade costs were high. The revenue from coffee planting and processing accounted for less than 10% of industrial chains total revenue, and majority of the profits were generated in the circulation and sales links. The electronic world trade platform initiated by Alibaba has carried out cooperation and innovation in customs clearance, payment, logistics, etc. with Malaysia, Ethiopia, Rwanda and other countries. With the help of the electronic world trade platform, sales through Tmall Global can effectively reduce trade costs, benefit local coffee farmers, and allow Chinese consumers to taste boutique coffee beans originating from Ethiopia at a more favorable price. At the same time, the rapid expansion of the Chinese market has also increased the willingness of Ethiopian coffee manufacturers to purchase fully automatic coffee roasting and packaging equipment from China, enabling them to produce roasted coffee beans locally, and further expand the market. In January 2022, the Ethiopian ambassador live-streamed the sale of coffee on a Chinese e-commerce platform, and four tons of Ethiopian coffee beans were sold out in five seconds. In recent years, the sales value of Ethiopian coffee beans in the Chinese market has grown rapidly at an annual rate of 27%. According to Chinese customs data, China imported 20,000 tons of raw coffee beans from Ethiopia from January to November 2023 alone. At the 2023 WIC, the project "Alibaba Digital County Partnering with International Trade Center to Support Digital Poverty Alleviation in Developing African Countries" was selected as one of the 2023 "Excellent Cases of Working Together to Build a Community with a Shared Future in Cyberspace". Pamela Kirk-Hamilton, Executive Director of the Untied Nations International Trade Center, said that the development of e-commerce is an important channel for SMEs to achieve growth, and the project provides new development opportunities for developing countries in Africa.

8.4 International Cooperation on Cyberspace Governance

Safeguarding cybersecurity is a shared responsibility of the international community. China staunchly upholds the international system centered around the UN, the international order grounded in international law, and the fundamental principles of international relations based on the objectives and principles of the UN Charter. Building on these foundations, China works to formulate internationally accepted rules for cyberspace. China deepens international cooperation in cybersecurity emergency response, and collaborates with the global community to enhance data security and personal information protection, jointly combating cybercrime and cyberterrorism.

8.4.1 International Collaboration on Internet Governance

China consistently adheres to the principles of sovereign equality, non-use or threat of force, and peaceful settlement of disputes established in the UN Charter, and respects the rights of all countries to independently choose their own path of network development, network management model, Internet public policy, and equal participation in international governance of cyberspace. China firmly believes that all countries, regardless of their size, strength, and wealth, are equal members and have the right to equally participate in the construction of international rules and order to ensure that the future development of cyberspace is jointly controlled by people of all countries. In September 2020, the "Position Paper of the People's Republic of China on the 75th Anniversary of the United Nations" was released, calling on the international community to strengthen dialogue and cooperation on the basis of mutual respect, equality and mutual benefit, use cyberspace to promote economic and social development, international peace and stability and human well-being, oppose cyber warfare and cyber arms race, and jointly build a peaceful, secure, open, cooperative, and orderly cyberspace.

China supports the UN as the main channel in the international governance of cyberspace. It supports the UN to formulate a global convention to combat cybercrime, and has jointly proposed and promoted the adoption of a resolution by the UN General Assembly to establish an ad hoc intergovernmental committee of experts, calling for the early conclusion of an authoritative and universal convention to provide a legal basis for the international community to cooperate in responding to the challenges of cybercrime. China attaches great importance to giving full play to the key role of the UN in responding to international information security threats and, in collaboration with other member states of the SCO, has submitted to the UN the draft update of the "International Code of Conduct for Information Security". China has expanded its cooperation with specialized institutions of the UN on cyber affairs, participated in the formulation of the United Nations Educational, Scientific and Cultural Organization's "Recommendations on the Ethics of Artificial Intelligence", and carried out extensive cooperation with the World Intellectual Property Organization (WIPO) in the fields of domain name rule-making and domain name dispute resolution.

In March 2017, China issued its first "International Strategy of Cooperation on Cyberspace", which comprehensively and systematically proposed China's propositions on promoting international exchanges and cooperation on cyberspace for the first time, sending a positive signal to the world that China is committed to the peaceful development and win–win cooperation in cyberspace. China actively participates in the formation of regional cyber governance rules. China signed the RCEP, which covers a high-level international e-commerce rules. China actively promotes the process of joining the "Comprehensive and Progressive Agreement for Trans-Pacific Partnership" and the "Digital Economy Partnership Agreement", and participates in the formulation of high-standard rules in the field of digital economy.

China conducts dialogue and exchanges with the US with an attitude of equality and mutual respect. China is committed to conducting dialogue and exchanges with the US in the field of Internet on the basis of respecting each other's core concerns and properly managing differences, creating a sound market environment for the development of enterprises from all over the world, including the US, in China, and promoting cooperation in the field of cyberspace between China and the US However, for some time, the US has adopted a wrong policy toward China, which has caused serious difficulties in China-US relations, but China will adhere to the principle of independence and unswervingly safeguard its national sovereignty, security and development interests in cyberspace.

China deepens high-level cooperation with Russia in the field of cyberspace. In 2015, China and Russia signed the "Agreement on Cooperation between the Government of the People's Republic of China and the Government of the Russian Federation in the Field of Ensuring International Information Security", which set the direction for cooperation between the two countries in the field of information security. In 2021, on the occasion of the 20th anniversary of the signing of the "Treaty of Good-Neighborliness and Friendly Cooperation between the People's Republic of China and the Russian Federation", China and Russia issued a joint statement, in which the two sides reiterated that they would consolidate bilateral and multilateral cooperation in the field of international information security and continue to promote the construction of a global international information security system based on the principle of preventing conflicts in information space and encouraging the peaceful use of information technology.

China is committed to promoting China-EU cooperation in the field of cybersecurity with an open and inclusive attitude. It jointly established the China-EU Digital Economy and Cybersecurity Expert Working Group with the European Commission and held several meetings; in 2012, it established the China-EU Cyber Working Group mechanism, and the two sides continuously strengthened dialogue and cooperation in the cyberspace under the framework of the working group; it jointly issued the 2019 China-Germany Internet Economy Dialogue Outcome Document with Germany. China co-hosted several China-UK Internet Roundtables with the UK and reached a number of cooperation consensuses in the fields of digital economy, cybersecurity, child online protection, data and artificial intelligence.

China strengthens cooperation in the field of cybersecurity with its neighbors and other developing countries. The China-ASEAN Information Port Forum has been successfully held for consecutive years, continuously promoting cooperation between China and ASEAN countries in the digital field, and establishing a China-ASEAN network affairs dialogue mechanism; a China-Japan-South Korea trilateral network consultation mechanism has been established; the China-South Korea Internet Roundtable was co-hosted with South Korea; the China-Africa Internet Development and Cooperation Forum has been held the "China-Africa Digital Innovation Partnership Program" has been proposed; China-South Africa New Media Roundtable, China-Tanzania (Tanzania) Network Culture Exchange Conference, China-Kenya (Kenya) Digital Economy Cooperation and Development Seminar and other activities have strengthened exchanges and cooperation between China and Africa in the fields of new media, network culture, digital economy, etc.; multiple online Silk Road conferences were held to carry out pragmatic cooperation with Arab countries in the fields of information infrastructure, cross-border e-commerce, smart cities, etc.

8.4.2 International Cooperation on Cybersecurity

Chinese enterprises, scientific research institutions and universities actively participate in the formulation of international standards in the fields of ICT and cybersecurity, and promote the development of emerging technologies. Chinese institutions continue to strengthen cooperation with standard organizations such as the International Organization for Standardization (ISO) and the International Electrotechnical Commission (IEC), as well as industry organizations such as the International Association for Automatic Identification and Mobile Technology (AIM Global) and the Data Identifier Management Committee (DIMC). China's National Technical Committee for Cybersecurity Standardization is responsible for the technical business work of the Information Security Subcommittee of the First Joint Technical Committee of ISO and IEC (ISO/IEC JTC1 SC27, referred to as "SC27"). Since 2004, it has participated in the SC27 International Cybersecurity Standardization Working Conference for many consecutive years, promoting a number of China's independently developed cybersecurity standards to be transformed into international standards. Many Chinese experts serve as conveners and liaison officers of SC27, contributing Chinese wisdom to the formulation of international cybersecurity standards.

China has established international channels in the field of cybersecurity emergency response. The National Computer Network Emergency Response Technical Team/Coordination Center (CNCERT/CC) engages in exchanges with major national computer emergency response organizations, government departments, international organizations and alliances, Internet service providers, domain name registries, academic institutions and other Internet-related companies and organizations around the world. As of 2021, it had established "CNCERT International Partnership" with 274 computer emergency response organizations in 81 countries and regions, and signed cybersecurity cooperation memoranda with 33 of them. China has also established the "China-ASEAN Cybersecurity Exchange and Training Center" with ASEAN to jointly enhance cybersecurity capabilities.

China promotes in-depth cybersecurity cooperation with specific regional organizations and countries. In 2017, the BRICS countries reached an agreement on the "BRICS Practical Cooperation Roadmap for Cybersecurity". In 2021, the SCO Information Security Experts Group unanimously adopted the "SCO Member States Cooperation Plan for Ensuring International Information Security 2022–2023". In 2021, China and Indonesia signed the "Memorandum of Understanding on the Development of Cybersecurity Capacity Building and Technical Cooperation". In 2022, China and Thailand signed the "Memorandum of Understanding on Cybersecurity Cooperation".

8.4.3 Cooperation Against Cybercrime

China has always supported international cooperation on combating cybercrime. China promoted the UN Intergovernmental Experts Group on Cybercrime to hold seven meetings from 2011 to 2021, and made important contributions to the adoption of relevant resolutions on launching the formulation of a UN global convention on combating cybercrime. China participated in the signing of important documents such as the "Statement of the Heads of the SCO Member States on Jointly Combating International Terrorism" under the framework of the SCO, jointly combating terrorism, including cyberterrorism, as well as separatism and extremism. China hosted and actively participated in a series of meetings of the BRICS Counter-Terrorism Working Group, introduced China's specific practices in combating cyberterrorism, and put forward suggestions for BRICS countries to strengthen cooperation and exchanges on cyberterrorism.

China strengthens international law enforcement and judicial cooperation in cybersecurity. China has reached a consensus on cooperation in the field of cybersecurity with many countries, and carried out in-depth and pragmatic cooperation in combating cyberterrorism and telecommunications and cyber fraud. In combating cyberterrorism, through various forms such as joint counter-terrorism exercises, joint border defense operations, police cooperation, and judicial assistance, China has continuously deepened exchanges and cooperation with relevant countries, jointly responded to threats and challenges, and jointly maintained world peace and regional stability. In combating telecommunications and cyber fraud, China has actively carried out international law enforcement and judicial cooperation, and jointly investigated and handled major cross-border cases with many countries, achieving significant results. From March to June 2022, under the framework of International Criminal

Police Organization (ICPO), China and other 75 member states jointly participated in the "Dawn Operation", arrested more than 2000 criminal suspects, intercepted more than 50 million USD of illegal funds, and effectively curbed transnational telecommunications and cyber fraud activities.

China joins hands with countries around the world to protect the online rights and interests of minors. China actively cooperates with international organizations such as the UN Children's Fund and the International Association of Internet Hotlines, as well as relevant departments of the UK, Germany, the United Arab Emirates and other countries to address the problem of online pornography of minors; it has joined the "WePROTECT Global Alliance to End Online Child Sexual Exploitation", and worked with more than 200 governments, enterprises and civil society organizations around the world to combat online child sexual exploitation and abuse, and create a safer online environment for children.

8.5 Summary

In the contemporary digital era, the development of the global cyberspace confronts numerous challenges. The digital divide is steadily expanding. Developing countries are lagging behind in informatization construction. Cybersecurity risks loom large, and digital trade rules call for urgent improvement. These issues significantly impede the sound and orderly development of global cyberspace, posing tangible threats to the cybersecurity and economic interests of all countries. It is, therefore, of utmost urgency for the international community to jointly address the challenges in cyberspace governance. Against this backdrop, promoting global digital economic and trade cooperation has become an irresistible trend.

China has consistently championed the concept of a "community with a shared future for humanity" and actively engaged in global cyberspace governance, offering Chinese wisdom and Chinese solutions to tackle the aforesaid problems. The Chinese government has put forward the crucial proposition of building a "community with a shared future in cyberspace". Through a series of high-profile events like the WIC, China has communicated its firm resolve and proactive initiatives in cyberspace governance to the world, effectively facilitating the translation of this proposition from vision into practice. China advocates the construction of communities of shared development, security, responsibility, and interests. The aim is to foster common development, coordinated security safeguarding, extensive governance participation, and equitable achievement-sharing among countries in cyberspace. This provides a comprehensive, systematic, and constructive framework for global cyberspace governance, guiding the global cyberspace to evolve steadily towards a more equitable, secure, and prosperous future.

In the field of international cooperation in the digital economy, China has exhibited its responsibility as a major country and played an indispensable role. Regarding information infrastructure construction, China is deeply involved in the management of Internet infrastructure resources and actively participates in international

ICT infrastructure construction projects, contributing to global connectivity. In the provision of global digital services, China offers high-quality digital services such as BDS and cloud services, providing efficient and reliable technical support to countries worldwide. Meanwhile, China actively participates in digital economic governance cooperation under multilateral mechanisms, vigorously promotes the development of cross-border e-commerce, and collaborates closely with numerous countries to achieve mutual benefit and win–win outcomes, effectively spurring the robust development of global digital economy and driving the economic growth and digital transformation of participating countries.

In the field of international cooperation in cyberspace governance, China also assumes a pivotal leading role. China has always strictly abided by the principles of the UN Charter, firmly supported the leading position of the UN in international cyberspace governance, and actively participated in the formulation of regional cyberspace governance rules. It conducts extensive dialogue and cooperation with other countries on the basis of equality, mutual trust, and cooperation. In the field of cybersecurity, China actively participates in the formulation of international standards and establishes international emergency response channels. In the fight against cybercrime, China vigorously promotes international cooperation and actively joins hands with other countries to protect the cyberspace rights and interests of minors. China's unremitting efforts have laid a solid foundation for maintaining the security and stability of global cyberspace and safeguarding the cyberspace rights and interests of all countries.

Over the years, China has adhered to the UN as the main channel, deeply participated in the formulation of international rules for the cyberspace, earnestly engaged in the construction of global information infrastructure, and made every endeavor to promote the healthy development of the global Internet. China also enables countries around the world to share China's development opportunities through various means, especially cross-border e-commerce. A series of plans proposed by China and its practical actions have injected strong impetus into the development of the global digital economy and cyberspace governance, which fully demonstrates China's responsibility as a major country, effectively promotes the peace, security, and prosperity of global cyberspace, and contributes China's strength to building a better digital world.

References

1 Basics of the ICT Industry

China Academy of Information and Communications Technology. (2020). *Introduction to the Digital Economy: Theory, Practice, and Strategy*. Posts & Telecom Press.

China Academy of Information and Communications Technology. (2022, November 7). *White Paper on the Global Digital Economy (2022)*. https://www.caict.ac.cn/kxyj/qwfb/bps/202212/P020221207397428021671.pdf

China Research Institute of Electronic Information Engineering Technology Development Strategy. (2017). *Research on the Development of China's Electronic Information Engineering Science and Technology (Comprehensive Volume)*. Science Press.

Freeman, C. (1994). The Economics of Technical Change. *Cambridge Journal of Economics, 5*, 463–514.

Gao, W., et al. (2019). *Research on Major Action Plans for the Development of the New Generation Information Industry*. Science Press.

National Bureau of Statistics of China.(2021, May 27). *Statistical Classification of Digital Economy and Its Core Industries (2021)*. https://www.gov.cn/gongbao/content/2021/content_5625996.htm

Negroponte, N. (1997). *Being Digital* (Hu Y. Haikou, Trans.). Hainan Publishing House.

Perez, C. (2003). *Technological Revolutions and Financial Capital: The Dynamics of Bubbles and Golden Ages*. Edward Elgar Publishing.

Sendler, U. (2014). *Industry 4.0: The Coming Fourth Industrial Revolution* (Deng Min, Li Xianmin, Trans.). China Machine Press.

United Nations Statistics Division, Department of Economic and Social Affairs. (2009). *International Standard Industrial Classification of All Economic Activities (ISIC), Rev.4*. United Nations. https://unstats.un.org/unsd/publication/SeriesM/seriesm_4rev_4.pdf

Wang, G. Y. (2015). *The Rise of Intelligence in 2049: Medium- and Long-Term Development Strategy for the New Generation of Information Technology Industry*. CITIC Press.

Wu, Y., & Zhu, M. (2014). *The New Generation of Information Technology Industry*. Shanghai Scientific and Technological Literature Publishing House.

Zhang, X. L. (2016). *Independence and Control: China's Power in Information Industry Innovation*. Tsinghua University Press.

Zhou, Y. B. (2017). *A Brief History of China's Information Industry*. Intellectual Property Publishing House.

W. Liu et al. (eds.), *Digital China*, Introduction to China's S&T Innovation,
https://doi.org/10.1007/978-981-95-6240-4

2 Overview of Digital China

Central Cyberspace Affairs Commission. (2021, December 28). *"14th Five-Year Plan" for National Informatization*. https://www.gov.cn/xinwen/2021-12/28/5664873/files/1760823a103e4d75ac681564fe481af4.pdf

China Academy of Information and Communications Technology. (2021, April 23). *China's Digital Economy Development White Paper (2021)*. https://www.caict.ac.cn/kxyj/qwfb/bps/202104/P020210424737615413306.pdf

China Academy of Information and Communications Technology. (2022). *2022 ICT In-depth Observation*. People's Posts and Telecommunications Press.

China Academy of Information and Communications Technology. (2022, July 8). *China Digital Economy Development Report (2022)*. https://www.caict.ac.cn/english/research/whitepapers/202208/P020220819505049573088.pdf

China Association of Communication Enterprises. (2022). *2021–2022 Analysis Report on the Development of China's Information and Communication Industry*. People's Posts and Telecommunications Press.

China Internet Network Information Center. (2022, February 25). *The 49th Statistical Report on China Internet Development*. https://www.cnnic.cn/NMediaFile/2023/0807/MAIN1691372884990HDTP1QOST8.pdf

Cyberspace Administration of China. (2021, July 2). *Digital China Development Report (2020)*. https://gxj.xm.gov.cn/xwzx/tzgg/202107/P020210712364847771519.pdf

Cyberspace Administration of China. (2022, August 2). *Digital China Development Report (2021)*. https://www.cac.gov.cn/2022-08/02/c_1661066515613920.htm

Forward Industry Research Institute. (2020, August 30). *2020 China Digital Economy Development Report*. https://www.idcode.org.cn/UploadFiles/20210426160807200.pdf

Future Industry Research Group of Institute of Industrial Economics of CASS. (2020). *China's New Infrastructure: Future Layout and Course of Action*. CITIC Press.

He, W., et al. (2022a). *Digital China: Insight into New Trends in Industrial Digitalization*. People's Posts and Telecommunications Press.

Huang, Q. F., et al. (2021). *Digital China*. CITIC Press.

Institute of Electronic Information of CCID. (2020, March 23). *New Infrastructure Development White Paper*. https://www.ccidgroup.com/info/1096/21559.htm

Institute of Party History and Literature of the Central Committee Party of China. (2021). *Xi Jinping's Discourses on Network Power*. Central Literature Press.

Liu, R. (2022). Characteristics, Problems and Future Direction of China's New-Generation Information Technology Industry Policies-Analysis Based on 2011–2020 Policy Texts. *Global Science and Technology Economic Outlook, 6*(37), 5–10.

Lv, B. F., et al. (2021). *Research on the Development Strategy of Network Power*. Anhui Education Press.

Ministry of Industry and Information Technology of the People's Republic of China. (2021, November 1). *"14th Five-Year Plan" Information and Communication Industry Development Plan*. Retrieved December 21, 2023, from http://big5.www.gov.cn/gate/big5/www.gov.cn/zhengce/zhengceku/2021-11/16/5651262/files/96989dadf83a4302895cd17cbeec6600.pdf

Ministry of Science and Technology of the People's Republic of China. (2019). *70 Years of China's Science and Technology Development*. Science and Technology Literature Press.

State Council of the People's Republic of China. (2021, December 12). *Circular of the State Council on the Issuance of the 14th Five-Year Plan for the Development of the Digital Economy: Guo Fa [2021] No. 29*. Retrieved December 21, 2023, from https://www.gov.cn/gongbao/content/2022/content_5671108.htm

State Council of the People's Republic of China. (2021, March 13). *Outline of the Fourteenth Five-Year Plan for the National Economic and Social Development of the People's Republic of China and the Visionary Goals for 2035*. Retrieved December 21, 2023, from https://www.gov.cn/xinwen/2021-03/13/content_5592681.htm

State Council of the People's Republic of China. (2023, February 27). *Overall Layout Plan for the Construction of Digital China*. Retrieved December 21, 2023, from https://www.gov.cn/xinwen/2023-02/27/content_5743484.htm

Wu, N. C. (2021). *Reading the New Infrastructure: Digital Technology Brings Opportunities for All*. Electronic Industry Press.

Yang, Y. M. (2019). A Review of 70 Years of Informatization Development in New China and the Summary of Experience. *Journal of Qiqihar University (Philosophy and Social Science Edition)*, (9), 36–40+44.

Ye, X. M. (2017). *History of E-commerce Development in China*. Shanxi Economic Press.

Yin, L. (2020a). *Digital Economy Development Report (2019–2020)*. Electronic Industry Press.

Zhang, L. (2020a). *Blue Book on the Development of China's Electronic Information Industry 2019–2020*. Electronic Industry Press.

Zhao, Y. (2021a). *Digital Economy Development Report (2020–2021)*. Electronic Industry Press.

3 ICT Innovations

CCID Consulting Co., Ltd., Digital Economy Branch of China Computer Users Association (Digital Economy Industry Alliance). (2020, September 22). *5G Industry Development White Paper (2020)*. Retrieved December 21, 2023, from https://www.ccidgroup.com/info/1096/21319.htm

China Academy of Information and Communications Technology. (2021, December 24). *Research Report on the Development and Application of Quantum Information Technology*. Retrieved December 21, 2023, from https://www.caict.ac.cn/kxyj/qwfb/bps/202112/P020211224561566573378.pdf

China Center for Information Industry Development, Policy Working Group of China Ultra HD Video Industry Alliance. (2021, May 9). *Ultra HD Video Industry Development White Paper (2021)*. Retrieved December 21, 2023, from https://docs.qq.com/pdf/DWUtCWm1LZ1dnV1Fn?

Cyberspace Administration of China. (2021, December 28). *National Informatization Plan for the 14th Five-Year Plan*. Retrieved December 21, 2023, from https://www.gov.cn/xinwen/2021-12/28/5664873/files/1760823a103e4d75ac681564fe481af4.pdf

GNSS and LBS Association of China. (2022, May 18). *2022 White Paper on the Development of China's Satellite Navigation and Positioning Service Industry*. Retrieved December 21, 2023, from https://fe.faisco.cn/pdfjs/web/viewer.html

Institute of Electronic Information of CCID. (2019, September 11). *Quantum Computing Development White Paper (2019)*. Retrieved December 21, 2023, from https://www.ccidgroup.com/info/1207/42117.htm

iResearch. (2022, September 1) *China Semiconductor IC Industry Research Report 2022*. Retrieved December 21, 2023, from https://report.iresearch.cn/report_pdf.aspx?id=4055

Ministry of Science and Technology of the People's Republic of China. List of National High-Tech Industrial Development Zones. Retrieved December 21, 2023, from https://www.most.gov.cn/zxgz/gxjscykfq/gxjsgxqml/

Ministry of Science and Technology of the People's Republic of China. (2019). *70 Years of China's Scientific and Technological Development*. Science and Technology Literature Press.

National Development and Reform Commission of the People's Republic of China. (2019, September). *Notice of the National Development and Reform Commission on Accelerating the Work Related to the Construction of Strategic Emerging Industry Clusters:Development and Reform High-Tech (2019)*, 1473.

Shao, S. H., Wu, C., & Huang, X. H. (2021). *The Road of Chinese Science and Technology—Information Volume—Intelligent Connection of Everything*. People's Posts and Telecommunications Press.

State Council of the People's Republic of China. (2021, March 13). *Outline of the Fourteenth Five-Year Plan for the National Economic and Social Development of the People's Republic of China and the Visionary Goals for 2035*. Retrieved December 21, 2023, https://www.gov.cn/xinwen/2021-03/13/content_5592681.htm

World Intellectual Property Organization. (2021, March 21). *World Intellectual Property Indicators 2021*. Retrieved December 21, 2023, from https://www.wipo.int/edocs/pubdocs/en/wipo_pub_941_2021.pdf

World Intellectual Property Organization. (2021, September 20). *Global Innovation Index 2021*(14th ed.). Retrieved December 21, 2023, from https://www.wipo.int/edocs/pubdocs/en/wipo_pub_gii_2021_exec.pdf

Yuanxi, Y. (2022). *The Road of Chinese Science and Technology-Aerospace Volume-BeiDou Navigation*. Science Press.

Yang, X. X., Chen, H. Y., et al. (2014). *History of Contemporary Chinese Science and Technology*. Intellectual Property Press.

Yang, Y. L. (2021). *China's Road of Science and Technology-Overview Volume-Scientific and Technological Power*. Science Press.

4 Digital Industrialization

Ai Rui Consulting. (2022, September 1). *2022 China Semiconductor IC Industry Research Report*. Retrieved December 21, 2023, from https://report.iresearch.cn/report_pdf.aspx?id=4055

China Academy of Information and Communications Technology. (2019, August 12). *Telematics White Paper (Net-connected Autonomous Driving Subsection)*. (2021–12–15) Retrieved December 21, 2023, from https://www.caict.ac.cn/kxyj/qwfb/bps/202012/P020201215382968589778.pdf

China Academy of Information and Communications Technology. (2019, August 12). *China Arithmetic Development Index White Paper*. Retrieved December 21, 2023, from https://www.caict.ac.cn/kxyj/qwfb/bps/202109/P020210918521091309950.pdf

China Academy of Information and Communications Technology. (2020, May 18). *2020 Digital China Industry Development Report (Information and Communication Industry Chapter)*. Retrieved December 21, 2023, from https://www.caict.ac.cn/kxyj/qwfb/bps/202005/P020200518608842463758.pdf

China Academy of Information and Communications Technology. (2020, December 15). *Internet of Things White Paper*. Retrieved December 21, 2023, from https://www.caict.ac.cn/kxyj/qwfb/bps/202012/P020201215379753410419.pdf

China Academy of Information and Communications Technology. (2021, September 1). *China Broadband Development White Paper (2021)*. Retrieved December 21, 2023, from https://www.caict.ac.cn/english/research/whitepapers/202112/P020211224524461536711.pdf

China Academy of Information and Communications Technology. (2021, July 27). *Cloud Computing White Paper*. Retrieved December 21, 2023, from https://www.caict.ac.cn/kxyj/qwfb/bps/202107/P020210727458966329996.pdf

China Academy of Information and Communications Technology. (2021, December 20). *Big Data White Paper*. Retrieved December 21, 2023, from https://www.caict.ac.cn/kxyj/qwfb/bps/202112/P020211220495261830486.pdf

China Academy of Information and Communications Technology. (2021, December 22). *Blockchain White Paper*. Retrieved December 21, 2023, from https://www.caict.ac.cn/kxyj/qwfb/bps/202112/P020211224394830046624.pdf

China Academy of Information and Communications Technology. (2022). *2022 ICT In-depth Observation*. People's Posts and Telecommunications Press.

China Academy of Information and Communications Technology. (2022, April). *Data Center White Paper (2022)*. Retrieved December 21, 2023, from https://www.caict.ac.cn/kxyj/qwfb/bps/202204/P020220422707354529853.pdf

China Academy of Information and Communications Technology. (2022, April 12). *Artificial Intelligence White Paper (2022)*. Retrieved December 21, 2023, from https://www.caict.ac.cn/kxyj/qwfb/bps/202204/P020220412613255124271.pdf

China Association of Communication Enterprises. (2022). *2021–2022 China's Information and Communication Industry Development Analysis Report*. People's Posts and Telecommunications Press.

China Center for Information Industry Development. (2020). *2019–2020 China Internet Industry Development Blue Book*. Electronic Industry Press.

China Federation of Electronics and Information Industry. (2022, February 15). *2021 China Electronic Information Industry Economic Operation Report*. Retrieved December 21, 2023, from https://lwzb.stats.gov.cn/pub/lwzb/tzgg/202205/W020220511403033781746.pdf

China Internet Network Information Center. (2022, February 25). *The 49th Statistical Report on the Development Status of the Internet in China*. Retrieved December 21, 2023, from https://www.cnnic.net.cn/NMediaFile/2023/0807/MAIN1691372884990HDTP1QOST8.pdf

Cyberspace Administration of China. (2021, July 2). *Digital China Development Report (2020)*. Retrieved December 21, 2023, from https://gxj.xm.gov.cn/xwzx/tzgg/202107/P020210712364847771519.pdf

Cyberspace Administration of China. (2022, August 2). *Digital China Development Report (2021)*. Retrieved December 21, 2023, from https://www.cac.gov.cn/2022-08/02/c_1661066515613920.htm

Forward Industry Research Institute. (2019, August 12). *2019 China Autonomous Driving Industry Development Research Report*. Retrieved December 21, 2023, from https://bg.qianzhan.com/report/detail/1908121708491526.html#read

Forward Industry Research Institute. (2019, August 19). *2019 Market Research Report on Internet of Things Industry*. Retrieved December 21, 2023, from https://bg.qianzhan.com/report/detail/1908191411410400.html#read

Forward Industry Research Institute. (2019, September 9). *China Short Video Industry Research Report 2019*. Retrieved December 21, 2023, from https://bg.qianzhan.com/report/detail/1909091648561802.html#read

Forward Industry Research Institute. (2019. November 4). *2009–2019 "Double 11" Shopping Festival Key Data Inventory and Development Trend Analysis Report*. Retrieved December 21, 2023, from https://bg.qianzhan.com/report/detail/1911041621381335.html#read

Forward Industry Research Institute. (2020, August 1). *2020 China Unmanned Economy Market Research Report*. Retrieved December 21, 2023, from https://bg.qianzhan.com/report/detail/2008111448446736.html#read

Forward Industry Research Institute. (2020, January 7). *2019 Blockchain Industry Status and Technology Application Analysis Report*. Retrieved December 21, 2023, from https://bg.qianzhan.com/report/detail/2001071621583572.html#read

Forward Industry Research Institute. (2020, May 12). *2020 China Live E-Commerce Research Report*. Retrieved December 21, 2023, from https://bg.qianzhan.com/report/detail/2005121533237188.html#read

Forward Industry Research Institute. (2020, November 5). *2020 China Semiconductor Materials Industry Development Report*. Retrieved December 21, 2023, from https://bg.qianzhan.com/report/detail/2011051011220492.html#read

Forward Industry Research Institute. (2020, October 22). *2020 China Semiconductor Equipment Industry Market Research Report*. Retrieved December 21, 2023, from https://bg.qianzhan.com/report/detail/2010221109417456.html#read

Forward Industry Research Institute. (2022, September 7). *Great Power Computing—Opportunity Outlook of East and West Computing in 2022*. Retrieved December 21, 2023, from https://bg.qianzhan.com/report/detail/2209071416019696.html#read

Institute of Cybersecurity of CCID, Blockchain Research Institute of CCID. (2021, March 18). *Blockchain + Digital Economy Development White Paper*. Retrieved December 21, 2023, from https://baijiahao.baidu.com/s?id=1696891122897630389&wfr=spider&for=pc

Institute of Radio Management of CCID. (2019, July 10). *White Paper on the Development of China's Satellite Communications Industry*. Retrieved December 21, 2023, from https://mp.weixin.qq.com/s?__biz=MjM5MzM4MjQwMA==&mid=2657675197&idx=1&sn=89438b17c0ca6fca19b98afef165e828&chksm=bd0baed28a7c27c4cd5fee1c6368726443f0d3e3a0b1e3c62362c8042decf21abc989df0289d&scene=310#wechat_redirect

iResearch. (2021, May 28). *2021 China Database Industry Research Report*. Retrieved December 21, 2023, from https://report.iresearch.cn/report_pdf.aspx?id=3787

iResearch. (2022, January 21). *2021 China Artificial Intelligence Industry Research Report*. Retrieved December 21, 2023, from https://report.iresearch.cn/report_pdf.aspx?id=3925

iResearch. (2022, January 29). *2022 China Internet of Things Industry Research Report*. Retrieved December 21, 2023, from https://report.iresearch.cn/report_pdf.aspx?id=3930

Ministry of Commerce of the People's Republic of China. (2022, November 16). *China E-Commerce Report 2021*. Retrieved December 21, 2023, from https://dzswgf.mofcom.gov.cn/news_attachments/e15a638b133d52134e0b1393fc13fc8a92bdd5ab.pdf

Ministry of Industry and Information Technology of the People's Republic of China. (2021, November 1). *"14th Five-Year Plan" Information and Communication Industry Development Plan*. Retrieved December 21, 2023, from https://www.gov.cn/zhengce/zhengceku/2021-11/16/5651262/files/96989dadf83a4302895cd17cbeec6600.pdf

Ministry of Industry and Information Technology of the People's Republic of China. (2021, March 24). *Circular of the Ministry of Industry and Information Technology on the Issuance of the Action Plan for the Cooperative Development of "Dual Gigabit" Networks (2021–2023): Ministry of Industry and Information Technology Communications [2021] No. 34*. Retrieved December 21, 2023, from https://www.gov.cn/zhengce/zhengceku/2021-03/25/content_5595693.htm

Ministry of Industry and Information Technology of the People's Republic of China. (2020, July 6). *"14th Five-Year Plan" Software and Information Technology Service Industry Development Plan*. Retrieved December 21, 2023, from https://www.gov.cn/zhengce/zhengceku/2021-12/01/5655205/files/a44b507d67c74591ad4f5e55b98c4518.pdf

Ministry of Industry and Information Technology of the People's Republic of China. (2022, July 6). *"14th Five-Year Plan" Big Data Industry Development Plan*. Retrieved December 21, 2023, from http://big5.www.gov.cn/gate/big5/www.gov.cn/zhengce/zhengceku/2021-11/30/5655089/files/d1db3abb2dff4c859ee49850b63b07e2.pdf.

Prospective Industry Research Institute. (2019, October 8). *2019 Artificial Intelligence Industry Status and Development Trend Report*. Retrieved December 21, 2023, from https://bg.qianzhan.com/report/detail/1910081709070618.html#read

QuestMobile. (2023, February 21). *2022 China Mobile Internet Annual Report*. Retrieved December 21, 2023, from https://mp.weixin.qq.com/s/A30Al0z7yH7Ow690mKgVcw

The State Council Information Office of the People's Republic of China. (2022, January 28). *2021 China's Spacefligh*. Retrieved December 21, 2023, from http://www.scio.gov.cn/zfbps/ndhf/2022n/202207/t20220704_130728.html

Yin, L. (2020). *2019–2020 Digital Economy Development Report*. Electronic Industry Press.

Zhang, L. (2020). *2019–2020 Blue Book on the Development of China's Electronic Information Industry*. Electronic Industry Press.

Zhao, Y. (2021). *2020–2021 Digital Economy Development Report*. Electronic Industry Press.

5 Industrial Digitalization

Beijing Software and Information Service Industry Association. (2021, August 2). *2021 Industrial Internet Practice Case Compilation.*

China Academy of Information and Communications Technology, Renmin University of China. (2021, December 24). *Research Report on China's Smart Agriculture Development - New Generation of Information Technology for Rural Revitalization.* Retrieved December 21, 2023, from https://pdf.dfcfw.com/pdf/H3_AP202201061538808867_1.pdf?1641482951000.pdf

China Academy of Information and Communications Technology. (2020, March 24). *Industrial Internet Industry and Economic Development Report (2020).* Retrieved December 21, 2023, from https://www.caict.ac.cn/kxyj/qwfb/bps/202003/P020200324455621419748.pdf

China Academy of Information and Communications Technology. (2023, April 21). *Smart Networked Vehicle Application Service Market Research Report (2023).* Retrieved December 21, 2023, from https://www.caict.ac.cn/kxyj/qwfb/ztbg/202304/P020230421601499344793.pdf

China Academy of Information and Communications Technology. China's (2021, December 27). *"5G+Industrial Internet" Development Report.* Retrieved December 21, 2023, from https://www.caict.ac.cn/kxyj/qwfb/bps/202112/P020211227607989259287.pdf

China Electronic Society. (2022, August 18).*China Robot Industry Development Report (2022).*

Forward Industry Research Institute. (2019, October 14). *2019 China Intelligent Manufacturing Development Status and Trend Analysis Report.* Retrieved December 21, 2023, from https://bg.qianzhan.com/report/detail/1910141608391476.html#read

GSM Association, China Academy of Information and Communications Technology. (2020, March 19). *China 5G Vertical Industry Application Cases (2020).*

He, W., Zuo, K. R., Zhang, D., et al. (2022). *Digital China: Insights into New Trends in Industrial Digitization.* People's Posts and Telecommunications Press.

Information Center of the Ministry of Agriculture and Rural Affirs of the People's Republic of China. (2022, June 7). *"14th Five-Year Plan" National Agricultural and Rural Informatization Development Plan.* Retrieved December 21, 2023, from https://www.caict.ac.cn/kxyj/qwfb/bps/202003/P020200324455621419748.pdf

Information Center of the Ministry of Agriculture and Rural Affirs of the People's Republic of China. (2023, March 1). *China Digital Countryside Development Report (2022).* Retrieved December 21, 2023, from https://www.cac.gov.cn/rootimages/uploadimg/1679309718522950/1679309718522950.pdf

Ministry of Industry and Information Technology of the People's Republic of China, et al. (2021, December 21). *Intelligent Manufacturing Development Plan for the 14th Five-Year Plan.* Retrieved December 21, 2023, from https://www.gov.cn/zhengce/zhengceku/2021-12/28/5664996/files/a22270cdb0504e518a7630fa318dbcd8.pdf

Ministry of Industry and Information Technology of the People's Republic of China. (2021, November 17). *Development Plan for the Deep Integration of Informatization and Industrialization in the 14th Five-Year Plan.* Retrieved December 21, 2023, from https://www.gov.cn/zhengce/zhengceku/2021-12/01/5655208/files/c09d992d37384268a73a201ef284909e.pdf.

Ministry of Industry and Information Technology of the People's Republic of China, et al. (2021, December 21). *Fourteenth Five-Year Plan for the Development of the Robot Industry.* Retrieved December 21, 2023, from https://www.gov.cn/zhengce/zhengceku/2021-12/28/5664988/files/7cee5d915efa463ab9e7be82228759fb.pdf

Ministry of Transportation of the People's Republic of China. (2021, September 23). *Action Program for New Infrastructure Construction in Transportation (2021–2025).* Retrieved December 21, 2023, from https://www.gov.cn/zhengce/zhengceku/2021-09/29/5639987/files/a2d1ca20cc0448cc9380c6f2f1f7c340.pdf

Ministry of Transportation of the People's Republic of China. (2022, January 12). *Development Plan for Digital Transportation in the 14th Five-Year Plan.* Retrieved December 21, 2023, from https://xxgk.mot.gov.cn/jigou/zhghs/202112/P020211222549694216970.pdf

National Energy Administration of the People's Republic of China. (2022, March 22). *14th Five-Year Plan for Modern Energy System*. Retrieved December 21, 2023, from https://www.nea.gov.cn/1310524241_16479412513081n.pdf
National Energy Administration of the People's Republic of China. (2023, April 20). *2022 Compilation of Typical Cases of 5G Application in Energy Sector*. https://www.nea.gov.cn/download/nylydxalhb.pdf
Publicity and Information Center of the Ministry of Industry and Information Technology. (2022). *5G Empowers All Industries to Interconnect and Lead the Future: China's 5G+Industrial Internet Application Demonstration Case Collection (2021)*. People's Posts and Telecommunications Press.

6 Digital Society

China Academy of Information and Communications Technology, Internet Society of China, China Communications Standards Association.(2023, Junuary 11). *Digital Twin City White Paper (2022)*. Retrieved December 21, 2023, from https://www.caict.ac.cn/kxyj/qwfb/bps/202301/P020230111662616392246.pdf
China Academy of Information and Communications Technology. (2022, November 23). *Research Report on the Development Trend and Construction Path of Digital Government*. Retrieved December 21, 2023, from https://www.caict.ac.cn/kxyj/qwfb/ztbg/202211/P020221123496269315741.pdf
China Academy of Information and Communications Technology. (2023, February 15). *Compilation of Typical Cases of Digital Government (2022)*. Retrieved December 21, 2023, from https://www.caict.ac.cn/kxyj/qwfb/ztbg/202302/P020230215518101160564.pdf
Forward Industry Research Institute. (2019, July 22). *2019 China Smart Education Industry Market Development and Trend Research Report*. Retrieved December 21, 2023, from https://bg.qianzhan.com/report/detail/1907221621522510.html#read
Forward Industry Research Institute. (2020, April 28). *2020 China Smart City Development Research Report*. Retrieved December 21, 2023, from https://bg.qianzhan.com/report/detail/2004281427056927.html#read
Information Center of the Ministry of Agriculture and Rural Affirs of the People's Republic of China. (2022, February 22). *"14th Five-Year Plan" National Agricultural and Rural Informatization Development Plan*. Retrieved December 21, 2023, from https://www.moa.gov.cn/zxfile/reader?file= http://www.moa.gov.cn/govpublic/SCYJJXXS/202203/P020220309588817315386.ofd
Information Center of the Ministry of Agriculture and Rural Affirs of the People's Republic of China. (2023, March 1). *China Digital Countryside Development Report (2022)*. Retrieved December 21, 2023, from https://www.cac.gov.cn/2023-03/01/c_1679309718486615.htm
iResearch. (2022, March 28). *China Primary and Secondary Education Informatization Industry Research Report 2022*. Retrieved December 21, 2023, from https://report.iresearch.cn/report_pdf.aspx?id=3959
iResearch. (2022, March 4). *China Healthcare Information Technology Industry Research Report 2022*. Retrieved December 21, 2023, from https://report.iresearch.cn/report_pdf.aspx?id=3942.
iResearch. (2023, Junuary 17). *China Digital Twin City Industry Research Report*. Retrieved December 21, 2023, from https://report.iresearch.cn/report_pdf.aspx?id=4133
Ministry of Commerce of the People's Republic of China. (2022, September 1). *China E-Commerce Report 2021*.
Ministry of Education of the People's Republic of China. (2016, June 7). *Circular of the Ministry of Education on the Issuance of the 13th Five-Year Plan for Educational Informatization: Education Technology [2016] No. 2*. Retrieved December 21, 2023, from https://www.gov.cn/gongbao/content/2016/content_5133005.htm

Ministry of Education of the People's Republic of China. (2018, April 13). *Circular of the Ministry of Education on the Issuance of the Education Informatization 2.0 Action Plan: Education Technology [2018] No. 6*. Retrieved December 21, 2023, from http://www.moe.gov.cn/srcsite/A16/s3342/201804/t20180425_334188.html

National Development and Reform Commission of the People's Republic of China. (2021, December 24). *"14th Five-Year Plan" to Promote National Government Informatization*. Retrieved December 21, 2023, from http://big5.www.gov.cn/gate/big5/www.gov.cn/zhengce/zhengceku/2022-01/06/5666746/files/cbff2937df654b44a04c6ef9d43549e9.pdf

National Health Commission of the People's Republic of China, National Administration of Traditional Chinese Medicine of the People's Republic of China, National Bureau of Disease Control and Prevention of the People's Republic of China. (2022, November 7). *Circular on the issuance of the "14th Five-Year Plan" on National Health Informatization: National Health Planning and Development [2022] No. 30*. Retrieved December 21, 2023, from http://www.nhc.gov.cn/guihuaxxs/s3585u/202211/49eb570ca79a42f688f9efac42e3c0f1.shtml

7 Cyberspace Governance

China Academy of Information and Communications Technology. (2020, September 16). *White Paper on China's Cybersecurity Industry (2020)*. Retrieved December 21, 2023, from https://www.caict.ac.cn/kxyj/qwfb/bps/202009/P020200916482039993423.pdf

China Academy of Information and Communications Technology. (2020, September 28). *White Paper on Artificial Intelligence Governance*. Retrieved December 21, 2023, from https://www.caict.ac.cn/kxyj/qwfb/bps/202009/P020200928368250504705.pdf

China Academy of Information and Communications Technology. (2020, March 1). *Research Report on Personal Information Protection in "Internet+Industry" (2020)*. Retrieved December 21, 2023, from https://www.caict.ac.cn/kxyj/qwfb/bps/202003/P020200302576687898634.pdf

China Academy of Information and Communications Technology. (2021, July 8). *White Paper on Trusted Artificial Intelligence*. Retrieved December 21, 2023, from https://www.caict.ac.cn/kxyj/qwfb/bps/202107/P020210709319866413974.pdf

China Academy of Information and Communications Technology. (2021, September 26). *"13th Five-Year Plan" White Paper on Online Copyright Governance in China*. Retrieved December 21, 2023, from https://www.caict.ac.cn/kxyj/qwfb/bps/202109/P020210926498663209246.pdf

China Academy of Information and Communications Technology. (2021, December 27). *Internet Law White Paper (2021)*. Retrieved December 21, 2023, from https://www.caict.ac.cn/kxyj/qwfb/bps/202112/P020211217543290402526.pdf

China Academy of Information and Communications Technology. (2021, November 19). *White Paper on the Governance of Personal Information Protection in Mobile Internet Applications (APP)*. Retrieved December 21, 2023, from https://www.caict.ac.cn/kxyj/qwfb/bps/202111/P020211119513519660276.pdf

China Academy of Information and Communications Technology. (2022, May 18). *Information Accessibility White Paper (2022)*. Retrieved December 21, 2023, from https://www.caict.ac.cn/kxyj/qwfb/bps/202205/P020220518510041281463.pdf

China Academy of Information and Communications Technology. (2023, January 12). *White Paper on Cyber Legislation (2022)*. Retrieved December 21, 2023, from https://www.caict.ac.cn/kxyj/qwfb/bps/202301/P020230114499859875704.pdf

China Internet Network Information Center. (2022, February 25). *The 49th Statistical Report on China Internet Development*. Retrieved December 21, 2023, from https://www.cnnic.cn/NMediaFile/2023/0807/MAIN1691372884990HDTP1QOST8.pdf

Chinese Academy of Cyberspace Studies. (2017). *China Internet 20 Years Development Report*. People's Publishing House.

Cyberspace Administration of China. (2022, August 2). *Digital China Development Report (2021)*. Retrieved December 21, 2023, from https://www.cac.gov.cn/2022-08/02/c_1661066515613920.htm
Institute of Party History and Literature of the Central Committee Party of China. (2021). *Xi Jinping's Discourses on Network Power*. Central Literature Press.
The State Council Information Office of the People's Republic of China. (2022). *Hand in Hand to Build a Community of Destiny in Cyberspace*. Foreign Languages Press.
The State Council Information Office of the People's Republic of China. (2023). *China's Internet Rule of Law in the New Era*. People's Publishing House.

8 International Digital Cooperation

China Academy of Information and Communications Technology. (2020, December 15). *White Paper on Global Digital Governance (2020)*. Retrieved December 21, 2023, from https://www.caict.ac.cn/kxyj/qwfb/bps/202012/P020201215465405492157.pdf
China Academy of Information and Communications Technology. (2020, December 16). *White Paper on the Development of Digital Trade (2020)*. Retrieved December 21, 2023, from https://www.caict.ac.cn/kxyj/qwfb/bps/202012/P020201216506475945126.pdf
China Academy of Information and Communications Technology. (2021, August 2). *Global Digital Economy White Paper - New Dawn of Recovery under the Shock of Epidemic*. Retrieved December 21, 2023, from https://www.caict.ac.cn/kxyj/qwfb/bps/202108/P020210913403798893557.pdf
China Academy of Information and Communications Technology. (2021, December 23). *White Paper on Global Digital Governance*. Retrieved December 21, 2023, from https://www.caict.ac.cn/kxyj/qwfb/bps/202112/P020211223383085909153.pdf
China Academy of Information and Communications Technology. (2022, December 7). *Global Digital Economy White Paper (2022)*. Retrieved December 21, 2023, from https://www.caict.ac.cn/kxyj/qwfb/bps/202212/P020221207397428021671.pdf
Chinese Academy of Cyberspace Studies. (2017b). *China Internet 20 Years Development Report*. People's Publishing House.
Forward Industry Research Institute. (2022, December 23). *2022 China and Global Digital Trade Development Trend Research Report*. Retrieved December 21, 2023, from https://bg.qianzhan.com/report/detail/2203231520200387.html#read
Institute of Party History and Literature of the Central Committee Party of China. (2021). *Xi Jinping's Discourses on Network Power*. Central Literature Press.
iResearch. (2021, February 28). *2021 China's New Cross-Border Export B2B E-commerce Industry Research Report*. Retrieved December 21, 2023, from https://report.iresearch.cn/report_pdf.aspx?id=3737
iResearch. (2022, June 28). *2022 China's Cross-Border E-commerce Service Industry Trend Report*. Retrieved December 21, 2023, from https://report.iresearch.cn/report_pdf.aspx?id=4014
The State Council Information Office of the People's Republic of China. (2022c). *Hand in Hand to Build a Community of Destiny in Cyberspace*. Foreign Languages Press.

The manufacturer's authorised representative in the EU is Springer Nature Customer Service Centre GmbH, Europaplatz 3, 69115 Heidelberg, Germany. If you have any concerns regarding our products, please contact ProductSafety@springernature.com

Printed and bound by CPI Group (UK) Ltd, Croydon, CR0 4YY
21/09/2026
02220012-0007